Diss. ETH No. 19798

# Depth, Crossings and Conflicts in Discrete Geometry

A dissertation submitted to
ETH Zurich

for the degree of
Doctor of Sciences

presented by
Marek Sulovský
Magistr, Charles University of Prague
born June 24, 1983
citizen of Czech Republic

accepted on the recommendation of
Prof. Dr. Emo Welzl, examiner
Prof. Dr. Boris Aronov, co-examiner
Dr. Uli Wagner, co-examiner

2011

Bibliografische Information der Deutschen Nationalbibliothek

Die Deutsche Nationalbibliothek verzeichnet diese Publikation in der Deutschen Nationalbibliografie; detaillierte bibliografische Daten sind im Internet über http://dnb.d-nb.de abrufbar.



ISBN 978-3-8325-3119-5

Logos Verlag Berlin GmbH
Comeniushof, Gubener Str. 47,
10243 Berlin
Tel.: +49 (0)30 42 85 10 90
Fax: +49 (0)30 42 85 10 92
INTERNET: http://www.logos-verlag.de

# Abstract

This thesis studies the following three combinatorial problems in discrete geometry: (i) centre points and depth of sets of points with respect to half-spaces or balls, (ii) crossings, their higher dimensional analogues and their significance in the context of $k$-sets, and (iii) conflict-free colourings of geometric hypergraphs.

First, we consider the circle containment problem introduced by Urrutia and Neumann-Lara in the late 1980s. The original question in the plane is the following: What is the largest number $f(n)$ such that in any set $P$ of $n \geq 2$ points in general position in the plane, there is a pair $\{a, b\} \subseteq P$ of disc depth at least $f(n)$, i.e. such that any disc $D$ containing $\{a, b\}$ has $|D \cap P| \geq f(n)$? We study a similar notion of depth with respect to various ranges (discs, balls, half-spaces) and improve the lower bounds on the higher dimensional variant of the problem – showing that every point set in $\mathbb{R}^d$ contains a subset of $\lfloor \frac{d+3}{2} \rfloor$ points of ball depth roughly $\frac{n}{d^2}$. This improves the previous lower bounds whose leading term was inversely exponentially dependent on the dimension $d$. As an intermediate step towards this improvement, we use random sampling and the bounds for the number of $(\leq k)$-sets to show that every point set in $\mathbb{R}^d$ has a small subset of high half-space depth. This has additional corollaries for several related geometric partitioning problems as well as for the structure of centre regions.

Next, we look into crossings, their generalisations, and crossing identities of $k$-edges and $k$-facets. As the main result in this direction, we extend an identity of Andrzejak et al. onto the sphere. The original identity relates the number of $k$-edges in a set of points in the plane to the number of crossings between them and gives an alternative way of proving an $O(nk^{1/3})$ upper bound on the number of $k$-edges in the plane. We show a similar identity on the sphere. Unlike in the plane, we need to consider an additional quantity measuring the "non-planarity" of the point set: its $g$-vector. There are two immediate consequences of the new identity: (i) It allows us to interpolate between the $k$-edge upper bound in the plane and the trivial (but tight) upper bound on the sphere, depending on the values $g_k$. (ii) Summing up our spherical identity around every point in a point set in $\mathbb{R}^3$ yields an identity in the three-dimensional space. However, the interaction of the terms in this identity is much less understood than in the planar one.

Finally, we investigate conflict-free colourings of geometric hypergraphs. This relatively new notion of colourings is motivated by frequency assignment in wireless networks: assignment of only few frequencies to base stations is necessary such that the clients can communicate without interferences. In the hypergraph setting, this is modelled as a so-called conflict-free colouring a hypergraph induced by a family of discs. We consider the list colouring variant of the problem, where we have restrictions on the possible colours of each vertex. Using a new idea of potential and weight updates, we are able to conflict-free list-colour hypergraphs from lists of size $O(\log n)$, as long as they are hereditarily properly colourable with constantly many colours. For instance, this is the case for all hypergraphs induced by planar regions of hereditarily linear union complexity. Furthermore, for many classes of geometric hypergraphs our bound matches what was already known for the less general non-list conflict-free colouring.

# Zusammenfassung


Diese Arbeit untersucht die folgenden drei Kombinatorischen Probleme. (i) Centerpunkte und Tiefe von Punktmengen in Bezug auf Halbräume oder Kugeln, (ii) Kreuzungen, ihre höherdimensionalen Analoga und deren Bedeutung im Kontext von $k$-Mengen und (iii) konfliktfreie Färbungen von geometrischen Hypergraphen.

Zunächst betrachten wir das Kreis-Umschliessungs-Problem, das in den späten 1980er von Urrutia und Neumann-Lara eingeführt wurde. Die ursprüngliche Frage in der Ebene ist die folgende: Was ist die grösste Zahl $f(n)$, so dass es in jeder Menge $P$ von $n \geq 2$ Punkten in allgemeiner Lage in der Ebene ein Paar $\{a, b\} \subseteq P$ gibt, das Kreisscheibentiefe mindestens $f(n)$ hat, d.h., für jede Kreisscheibe $D$, die $\{a, b\}$ beinhaltet, gilt, dass $|D \cap P| \geq f(n)$? Wir untersuchen einen ähnlichen Begriff von Tiefe in Bezug auf verschiedene Regionen (Kreisscheiben, Kugeln, Halbräume) und verbessern die unteren Schranken für die höherdimensionalen Variante des Problems — d.h., wir zeigen, dass jede Punktmenge in $\mathbb{R}^d$ eine Teilmenge von $\lfloor \frac{d+3}{2} \rfloor$ Punkte enthält, die Kugeltiefe von etwa $\frac{n}{d^2}$ hat. Dies verbessert die vorherigen unteren Schranken, in denen der führende Term umgekehrt-exponentiell Dimension abhängig. Als Zwischenschritt dieser Verbesserung verwenden wir eine zufällige Stichprobe und obere Schranken für $(\leq k)$-Mengen, um zu zeigen, dass jede Punktemenge in $\mathbb{R}^d$ eine kleine Teilmenge mit hoher Halbraum-Tiefe hat. Dies hat zusätzlich Folgen für mehrere verwandte geometrische Partitionierungsprobleme sowie für die Struktur der Centerpunktsregionen.

Als Nächstes betrachten wir Kreuzungen, deren Verallgemeinerungen und Kreuzungsidentitäten von $k$-Kanten und $k$-Facetten. Als wichtigstes Ergebnis in dieser Richtung erweitern wir eine Gleichung von Andrzejak et al. auf die 2-Sphäre. Die ursprüngliche Gleichung betrifft die Anzahl der $k$-Kanten in einer Menge von Punkten in der Ebene und die Zahl der Kreuzungen zwischen ihnen und gibt einen alternativen Beweis für die obere Schranke von $O(nk^{1/3})$ für die Anzahl der $k$-Kanten in der Ebene. Wir zeigen eine ähnliche Gleichung auf der Sphäre. Anders als in der Ebene brauchen wir eine zusätzliche Grösse zur Messung der "Nichtplanarität" der Punktmenge: den $g$-Vektor der Punktmenge. Die neue Gleichung hat zwei unmittelbare Folgen: (i) sie erlaubt uns, zwischen der oberen Schranke für $k$-Kanten in der Ebene und der trivialen (aber scharfen) obere Schranke auf der 2-Sphäre zu interpolieren, abhängig von den Werten $g_k$. (ii) Indem wir unserer sphärischen Gleichung über alle Punkte in einer Punktmenge in $\mathbb{R}^3$ aufsumieren, erhalten wir eine Gleichung im dreidimensionalen Raum, wobei in diesem Fall die Wechselwirkung zwischen den Termen in der Gleichung bisher noch weitaus klar ist als im planaren Fall.

Schliesslich untersuchen wir konfliktfreie Färbungen von geometrischen Hypergraphen. Dieser relativ neue Begriff von Färbungen ist durch Frequenzzuteilungen in drahtlosen Netzwerken motiviert: die Zuweisung von nur wenigen Frequenzen an Basisstationen ist notwendig, damit die Kunden interferenzfrei kommunizieren können. Im Rahmen von Hypergraphen entspricht dies einer konfliktfreien Färbung eines Hypergraphen, der durch eine Familie von Kreisscheiben induziert ist. Wir betrachten die Farblisten-Variante des Problems, bei dem die möglichen Farben einer Ecke beschränkt sind. Unter Verwendung einer neuen Idee, die auf Potenzialen und Gewichtaktualisierungen aufbaut, sind wir in der Lage, Hypergraphen mit Farblisten der Grösse $O(\log n)$ listenkonform zu färben, solange sie erblich mit konstant vielen Farben färbbar sind. Dies ist beispielsweise für alle Hypergraphen der Fall, die von planaren Regionen mit erblich linearer Vereinigungskomplexität induziert werden. Des weiteren stimmen die Schranken für viele Klassen von geometrischen Hypergraphen mit dem überein, was bereits für den weniger allgemeinen Fall von konfliktfreien Färbungen ohne Farblisten bekannt war.

# Acknowledgements

First and foremost, I owe profound gratitude to my advisor, Uli Wagner, for his guidance, discussions, ideas, patience, encouragements and endless supply of optimism.

I also feel indebted to Emo Welzl for giving me the opportunity to be part of his group. It has been an amazing and very stimulating environment for working.

Next, I want to thank all my collaborators and co-authors: Uli Wagner for opening the world of $k$-sets to me and always having interesting research problems; Shakhar Smorodinsky and Panos Cheilaris for making me a discrete-geometric equivalent of a painter; Martin Jaggi for cracking the matrix with me. It has been pleasure to work with all of you!

My sincere thanks go to Boris Aronov for reviewing this thesis and his helpful remarks as well as for coming all the way from New York to my final exam.

I would like to thank all the current and former Gremos, whom I had the pleasure to meet: Andrea Francke, Andrea Salow, Andreas Razen, Anna Gundert, Bernd Gärtner, Dieter Mitsche, Dominik Scheder, Emo Welzl, Eva Schuberth, Floris Tschurr, Franziska Hefti, Gabriel Nivasch, Heidi Gebauer, József Solymosi, Leo Rüst, Martin Jaggi, Michael Hoffmann, Miloš Stojaković, Patrick Traxler, Philipp Zumstein, Robert Berke, Robin Moser, Sebastian Stich, Tibor Szabó, Timon Hertli, Tobias Christ, Uli Wagner, and Yves Brise. I also want to thank all the other colleagues whom I had the opportunity to get to know better, one way or another: Christian Müller, Christoph Krautz, Dan Hefetz, Florian Jug, Ilan Karpas, Jan Foniok, Justus Schwartz, Konstantinos Panagiotou, Panagiotis Cheilaris, Rastislav Šrámek, Rom Pinchasi, Yann Disser, and Yelena Yuditsky.

In particular, I am grateful to Andrea for the healthy dose of decadence in Paris; Andreas for all his hilarious stories; Anna for sharing some of her green views on the world with us; Christian, Martin and Sebastian for all the fun at the workshop in Paris; Dominik, Kosta and Philipp for taking me for occasional drinks in Zürich's bars, where I got most of the German I have managed to master by now; Heidi for always being politically correct; Justus and Martin for being my climbing–mentors; Lena for being a guide and a friend to me in Be'er Sheva; Leo for introducing me to the red and blue guys in H52; Michael for having time for a game of töggeli whenever I needed a break from all the maths; Robin for unconsciously being the snobbery teacher I always needed; Rom for a great time in Haifa and around; Shakhar for the chance to have fun time over there in the desert; and Tobi and Yves for sharing the first steps through the jungle of doctoral research with me.

Last but not least, I am grateful to my family and my best friends. Without their support, it would only have been half the fun.

# Contents

# 0
# Introduction

This text is the result of several years' work on problems in discrete geometry throughout my doctoral studies. One might ask a philosophical question: what is discrete geometry, the topic of this work? Let me start with the biggest cliché of all and quote the description of this field from Wikipedia.

> "Discrete geometry and combinatorial geometry are branches of geometry that study combinatorial properties and constructive methods of discrete geometric objects. Most questions in discrete geometry involve finite or discrete sets of basic geometric objects, such as points, lines, planes, circles, spheres, polygons, and so forth. The subject focuses on the combinatorial properties of these objects, such as how they intersect one another, or how they may be arranged to cover a larger object."

Questions in discrete geometry usually involve finite sets of objects such as points, lines, hyperplanes, half-spaces, etc. and for some combinatorial relations between them. One of the classical problems in the field is the distinct distances problem: what is the minimum number of distinct distances between $n$ points in the plane? Another typical example is the $k$-set problem: what is the maximum number of $k$ point subsets of an $n$ point set, which are linearly separable from the rest?

The latter one was, to some extent, a topic of interest and inspiration for this work, especially due to a rich variety of methods used or developed for the problem itself.

## 0.1 Thesis in a nutshell

The three main areas discussed in this text are *crossings* in $\mathbb{R}^3$, the *circle containment* problem and *conflict-free colourings.* The glue that binds the first two problems together is the $k$-set problem mentioned earlier, whereas the third problem stands out of the group although it deals with similar objects as the second one – discs and pseudo-discs.

As we will refer to the $k$-set problem in this introduction several times, we introduce the problem before giving a more detailed overview of the topics. After the description of the $k$-set problem we will briefly discuss each of the three topics investigated in this work.

Consider a set $P$ of $n$ points in $d$-dimensional Euclidean space. A *$k$-set* of $P$ is a subset $A \subseteq P$ of cardinality $|A| = k$ for which there is a hyperplane $h$ strictly separating $A$ from $P \setminus A$.

The question of determining the maximum number $a_k^d(n)$ of $k$-sets in a set of $n$ points in $\mathbb{R}^d$ is known as the *$k$-set problem.* Determining the asymptotic behaviour of $a_k^d(n)$ when $d$ is fixed is considered one of the most difficult questions in discrete geometry. Despite decades of efforts of numerous researchers, the gap between the upper and lower bound remains enormous.

A notion which is often considered instead of $k$-sets is that of a $k$-facet. For a set $P \subseteq \mathbb{R}^d$ of $n$ points in general position (that is, every subset of cardinality at most $d+1$ is affinely independent) consider an oriented simplex $\sigma$ spanned by $d-1$ points of $P$. The affine hull of $\sigma$ is a hyperplane which splits the space into a positive and negative half-space. The simplex $\sigma$ is called a $k$-facet of $P$ if the positive half-space contains exactly $k$ points of $P$. The maximum number of $k$-facets in a set of $n$ points in $\mathbb{R}^d$ is denoted by $e_k^d(n)$. The numbers $e_k^d(n)$ and $a_k^d(n)$ are asymptotically equivalent. Due to this fact, many upper and lower bound proofs rather aim at the $k$-facets rather than $k$-sets due to their somehow more concrete structure.

Although the numbers of $k$-facets and $k$-sets are very difficult to estimate, there is a closely related notion which has been treated with much more success. A set $A \subseteq P$ is called $\leq k$-facet of $P$ if $A$ is an $i$-facet of $P$ for some $i \leq k$. The maximum number of $\leq k$-facets in a set of $n$ points in $\mathbb{R}^d$ is denoted by $e_{\leq k}^d(n)$. It might sound surprising to some that this quantity is much easier to control but indeed, there are asymptotically tight upper bounds for it. Estimating

$e^d_{\leq k}(n)$ was one of the first problems to which Clarkson and Shor applied their random sampling method [CS89].

### 0.1.1 Circle containment and depth

Let $P$ be a set of $n$ points in $\mathbb{R}^d$. We say that $S \subseteq \mathbb{R}^d$ has a *ball depth* (*half-space depth*, respectively) at least $k$ if every ball (half-space, respectively) containing $S$ contains at least $k$ points of $P$. A point $x \in \mathbb{R}^d$ of half-space depth $k$ is called a $k$-centre point of $P$.

Neumann-Lara and Urrutia [NLU88] raised the following question: given a set $P$ of $n$ points in the plane (in general position, i.e. no quadruple on a common circle), is there always a pair of points $\{a, b\} \subseteq P$ of disk depth at least $c \cdot n + o(n)$ (asymptotically; for some $c > 0$). They showed that the answer is yes and proved a lower bound $c \geq 1/60$ which has been improved several times to the current record [EHSS89] of

$$c \geq \frac{1}{2} - \sqrt{\frac{1}{12}} \approx \frac{1}{4.73}.$$

One of the papers improving the lower bound, by Bárány et al. [BSSU89], also considered a higher-dimensional variant of the problem. In dimension $d \geq 3$, pairs of points are no longer sufficient and one needs to consider subsets of size $\lfloor \frac{d+3}{2} \rfloor$. They showed that for every $d$, there is a function $f_d(n) \in \Theta(n)$ such that for every set $P$ of $n$ points in $\mathbb{R}^d$, there is a subset $X \subseteq P$ of size at most $\lfloor \frac{d+3}{2} \rfloor$ such that every ball containing $X$ also contains at least $f_d(n)$ points of $P$. The dependence of their lower bound is inversely exponential in the dimension.

In Part I, we improve their lower bound in higher dimensions (that is, $d \geq 4$). It turns out that the earlier mentioned random sampling technique of Clarkson and Shor together with precise upper bounds on the values $e^d_{\leq k}(n)$ can be used to give a lower bound for the ball depth problem with much more favourable dependence on the dimension. Along the way, we obtain structural properties of centre regions of points and several corollaries of these properties.

### 0.1.2 Crossings and $k$-edge structure

Let us have a closer look at upper and lower bounds on the quantities $a^d_k(n)$ defined earlier. There are only very few structural properties of $k$-facets that have been successfully used in the upper bound proofs. One common theme in $d = 2, 3, 4$ is bounding the number of crossing between certain objects related to the $k$-facets as an intermediate step.

The upper bound in the plane [Dey98] in an example which follows this theme beautifully: the main idea is to give an upper bound (in terms of the number of points $n$) on the number of crossings between $k$-edges (this is what $k$-facets are called in the plane); on the other hand, the crossing lemma [ACNS82, Lei84] gives a lower bound on the number of $k$-edge crossings (in terms of $n$ and the number of $k$-edges) and combining these two yields the desired upper bound on the number of $k$-edges. An alternative proof [AAHP+98] of the upper bound shows, that the $k$-edges and their crossings obey a certain identity.

In Part II, we prove an identity of a similar spirit for $k$-facets in $\mathbb{R}^3$, which we believe improves our understanding of their structure. As a key step along the way, we first extend the planar identity to point sets on the 2-sphere. A new ingredient compared to the planar case, is the winding number of $k$-facets around a given point in 3-space, as introduced by Lee [Lee91] and Welzl [Wel01]. Additionally, we present several thoughts about possible higher-dimensional notions of crossings.

### 0.1.3 Conflict-free colourings

A hypergraph $H = (V, \mathcal{E}), \mathcal{E} \subseteq 2^V$ is a generalization of a graph and there are many natural hypergraphs of a geometric nature. A notion of graph colouring can be generalised for hypergraphs in various ways and there are several definitions of hypergraph colourings with different theoretical or practical motivation. One of the recent ones, motivated by frequency assignment in cellular networks, is a notion of conflict-free colouring: a mapping $C : V \to \mathbb{Z}_0^+$ is called a *conflict-free colouring* if every hyperedge $S \in \mathcal{E}$ contains a vertex $v$ of a unique colour within $S$, i.e. $\forall u \in S \setminus \{v\} : C(v) \neq C(u)$.

The research of conflict-free colouring was initiated in the works [ELRS03] and [Smo03] and was further studied in many settings. The original motivation for the problems is completely geometric (colouring hypergraphs induced by families of discs corresponds to assigning frequencies to some antennae) and geometry also serves as the source of most studied examples of hypergraphs.

In Part III, we consider the list variant of conflict-free colouring (i.e. each vertex can only be coloured from its own list of admissible colours). We extend a general conflict-free colouring framework for hereditarily $k$-colourable hypergraphs and show that a framework with the same guarantees on the number of colours (sizes of the lists, in our case) also exists for conflict-free colouring from arbitrary lists. The result implies, in particular, that a hypergraph induced by a family of $n$ discs (or pseudo-discs) can be conflict-free coloured from lists of size logarithmic in $n$ (what exactly the hypergraph is will be precisely defined later).

## 0.2 Terminology and notation

This text assumes some basic background in graph theory, linear algebra and convex geometry. Nevertheless, we quickly summarise the notation and some basic geometric theorems and facts which are a prerequisite for further reading. A reader familiar with discrete mathematics and discrete geometry will be accustomed to most of the notation used throughout the book. This section will review some of the most basic notions which appear in the later chapters and an experienced reader may skip this section entirely. The not so common or problem-specific terminology will be introduced in the later sections when necessary.

**Numbers and sets.** The set of non-negative integers is denoted by $\mathbb{Z}_0^+$, the set of positive natural numbers $\mathbb{N}$, the set of all integers $\mathbb{Z}$, the set of real numbers $\mathbb{R}$ and the set of non-negative real numbers $\mathbb{R}_0^+$. For a real number $x \in \mathbb{R}$, the largest integer smaller or equal to $x$ is denoted by $\lfloor x \rfloor$ and the smallest integer larger or equal to $x$ by $\lceil x \rceil$. The *discrete interval* $\{a, a+1, \dots, b-1, b\}$ for $a, b \in \mathbb{Z}$ is denoted by $[a, b]$. We write $[n]$ as a shorthand for $[1, n]$. The set of all subsets of $S$ is called the *power set of* $S$ and is denoted by $2^S$ and a set of all subsets of $S$ of cardinality $k$ is denoted by $\binom{S}{k}$. The symbols $\subseteq$ and $\subset$ denote non-strict and strict inclusion, respectively. Let $\mathcal{F}$ be a family of sets. We write $\bigcup \mathcal{F}$ as a shorthand for $\bigcup_{A \in \mathcal{F}} A$ and similarly $\bigcap \mathcal{F}$ for $\bigcap_{A \in \mathcal{F}} A$.

**Mappings.** Let $f : X \to Y$ be a mapping. The image of a set $A \subseteq X$ is denoted by $f(A) := \{f(a) \colon a \in A\}$. The pre-image of a set $B \subseteq Y$ is denoted by $f^{-1}(B) := \{x \in X \colon f(x) \in A\}$. For a single element $y \in Y$ we write $f^{-1}(y)$ as a shorthand for $f^{-1}(\{y\})$. For an injective mapping, the inverse mapping is denoted by $f^{-1} : Y \to X$ as well. A restriction of $f$ to a set $A \subseteq X$ is the mapping $f|_A : A \to Y$ defined as $f|_A(a) = f(a)$ for every $a \in A$. At times we will abuse the notation and use $f(A)$ as a shorthand for $\sum_{a \in A} f(a)$ but this will be mentioned explicitly.

**Big-Oh notation.** We use the $O$-notation for the order of magnitude of functions at $\infty$. For two functions $f, g : \mathbb{N} \to \mathbb{R}_0^+$ we say $f \in O(g)$ if there is $n_0 \in \mathbb{N}$ and $c \in \mathbb{R}_0^+$ such that for every $n \in \mathbb{N}, n \geq n_0$ we have $f(n) \leq c \cdot g(n)$. Similarly, we write $f \in \Omega(g)$ if $g \in O(f)$ and $f \in \Theta(g)$ if $f \in O(g)$ and $f \in \Omega(g)$. When we want to ignore polylogarithmic factors, we use the $\tilde{O}$ notation: we write $f \in \tilde{O}(g)$ if there is $k \in \mathbb{N}$ such that $f \in O(g \cdot \log^k(n))$. When there are more variables and the usual $O$ notation might lead to a confusions, we write the variables treated as non-constants in the subscript of $O$, e.g. $f(m,n) \in O_n(g(m,n))$ denotes that for any fixed constant $m = c$ it holds that $f(c,n) \in O(g(c,n))$.

**Euclidean spaces.** The $d$ dimensional Euclidean space equipped with the usual $\ell_2$ norm is denoted by $\mathbb{R}^d$. The norm of a point $x$ is denoted by $\|x\|$ and the distance between $x$ and $y$ is denoted by $\operatorname{dist}(x,y) := \|x-y\|$. For the scalar product of $x$ and $y$, matrix notation $x^T y$ will be used. The coordinates of a point $x \in \mathbb{R}^d$ are denoted by $x_1, \ldots, x_d$. A translation of a set $A \subseteq \mathbb{R}^d$ by vector $x \in \mathbb{R}^d$ is denoted by $A + x := \{a + x \colon a \in A\}$. We write $\mathbb{S}^d$ for the $d$ dimensional *unit sphere*, i.e., $\mathbb{S}^d = \{x \in \mathbb{R}^{d+1} \colon \|x\| = 1\}$.

The smallest linear subspace containing a set $U \subseteq \mathbb{R}^d$ is called *linear span* and denoted by $\operatorname{span}(U)$. An *affine span* of a set $U \subseteq \mathbb{R}^d$ is the set $\operatorname{aff}(U) := x + \operatorname{span}(U - x)$ for $x \in U$. This is independent of the choice of $x$. An *affine subspace* $A$ of $\mathbb{R}^d$ is a set with $A = \operatorname{aff}(A)$. The *dimension* $\dim(A)$ of an affine subspace $A$ is one less than the size of a smallest subset $X \subseteq A$ such that $A = \operatorname{aff}(X)$ and its codimension is $\operatorname{codim}(A) := d - \dim(A)$. For an arbitrary set $X \subseteq \mathbb{R}^d$ we write $\dim(X) := \dim(\operatorname{aff}(X))$ and $\operatorname{codim}(X) := \operatorname{codim}(\operatorname{aff}(X))$. A *hyperplane* is an affine subspace of codimension 1.

A hyperplane $h$ can be written in the form $\{x \in \mathbb{R}^d \colon a^T x + b = 0\}$ for some point $a \in \mathbb{R}^d \setminus \{\mathbf{0}\}$ and a scalar $b \in \mathbb{R}$. It is called *non-vertical* if $a_d \neq 0$ in which case we assume $a_d > 0$ for the rest of this paragraph. The hyperplane $h$ defines two regions $\{x \in \mathbb{R}^d \colon a^T x + b \leq 0\}$ and $\{x \in \mathbb{R}^d \colon a^T x + b \geq 0\}$ called (closed) *half-spaces*. For non-vertical hyperplanes, the former one is called *lower* and the latter one *upper*. Open half-spaces are defined similarly with strict inequalities.

On occasions, we will speak of an orthogonal complement of an affine subspace $A$. Unlike for linear subspaces this is not well defined. The affine subspace* $x + (A - x)^\perp$ will be called *an orthogonal complement* of $A$ for any choice of the vector $x \in A$ and by $A^\perp$ we will denote some particular orthogonal complement (e.g. the one containing $\mathbf{0}$). For our purposes, the exact choice will not matter.

The *convex hull* of a set $X \subseteq \mathbb{R}^d$ is denoted by $\operatorname{conv}(X)$. The set $X$ is in (non-strictly) *convex position* if no point of $X$ lies in the interior of $\operatorname{conv}(X)$. The *diameter* of a set $X \subseteq \mathbb{R}^d$ is the supremum of distances between points of $X$, i.e. $\operatorname{diam}(X) := \sup_{u,v \in X} \|u - v\|$.

*Notice that the second term is a linear subspace and therefore, we can take its orthogonal complement in the usual sense.

**Open and closed sets.** A set $X \subseteq \mathbb{R}^d$ is called *closed* if every convergent sequence $(x_i)_{i=1}^{\infty}$ of points in $X$ also has its limit inside $X$. An *$\varepsilon$-neighbourhood* of a point $x$ is the set $\mathrm{N}_\varepsilon(x) := \{y \in \mathbb{R}^d \colon \|y - x\| < \varepsilon\}$. The *complement* of a set $X$ is the set $\mathbb{R}^d \setminus X$ and the set $X$ is called *open*, if its complement is closed. The *closure* of $X$ is the intersection of all closed sets containing $X$ and is denoted by $\overline{X}$. The *interior* of $X$, denoted by $\mathrm{int}(X)$, is the set of all points of $X$ which are contained in some open ball in $X$. The *boundary* of $X$ is the set $\partial X := \overline{X} \setminus \mathrm{int}(X)$. If $X$ lies in a lower-dimensional affine subspace then $\mathrm{int}(S) = \emptyset$. In this case it is sometimes more desirable to consider the notion of *relative interior* of $S$:

$$\mathrm{relint}(S) := \{x \in S \colon \exists \varepsilon > 0, \mathrm{N}_\varepsilon(x) \cap \mathrm{aff}(S) \subseteq S\}.$$

**Polytopes and polyhedra** A set $X \subseteq \mathbb{R}^d$ is called a *convex polyhedron* (or *polyhedral*) if it can be written as an intersection of finitely many half-spaces and it is called a *convex polytope* if it can be written as a convex hull of finitely many points. Every polytope is a polyhedron but not necessarily the other way round. Let $X$ be a convex polyhedron: a hyperplane $h$ is called a *supporting hyperplane* if it intersects $X$ and the whole polyhedron lies in one of the closed half-spaces defined by $h$. Every such possible intersection of a supporting hyperplane $h$ with $X$ is called a *face* $F$ of $X$. Faces of dimension $0, 1, d-1$ are called *vertices*, *edges* and *facets* of $X$, respectively. The set of vertices of a polytope $X$ is denoted by $V(X)$. A hyperplane $h$ whose intersection with $X$ is a facet is called a *facet-supporting hyperplane.*

**Regions in the plane.** A *Jordan curve* or a *simple closed curve* in the plane is an image of an injective continuous map $\phi : \mathbb{S}^1 \to \mathbb{R}^2$. In another words, it is a non-self-intersecting continuous loop in the plane. The Jordan curve theorem asserts that every Jordan curve divides the plane into *the interior* bounded by the curve and *the exterior.* A *Jordan region* is a region of the plane, whose boundary is a Jordan curve. A family $\mathcal{P}$ of simple closed Jordan regions is called a *family of pseudo-discs*, if for every pair $A, B \in \mathcal{P}$ the boundaries of $A$ and $B$ intersect in at most two points. An area of a planar region $X$ is denoted by $\mathrm{area}(X)$.

**Probability.** Probability of an event $A$ is denoted $\mathbf{Pr}[A]$ and the expectation of a random variable $X$ is denoted by $\mathbf{E}[X]$.

**Davenport-Schinzel sequences.** A *Davenport-Schinzel sequence of order $s$ with $n$ symbols* is a sequence $(a_i)_{i=1}^{l}$ of symbols $a_i \in [n]$ such that no two consecutive symbols are the same and it does not contain any sequence $aba \dots ba$ of length $s+2$ for $a, b \in [n], a \neq b$ as a subsequence. The maximum length $l$ of a Davenport-Schinzel sequence of order $s$ with $n$ symbols is denoted by $\lambda_s(n)$. This number is only slightly superlinear in $n$ (for $s$ fixed) and the inverse Ackermann function enters the scene for the bounds (see e.g. [Niv10]).

## 0.3 Preliminaries

As it is common in many geometry papers and books we will heavily use some classical geometric results describing intersection patterns of sets.

The first theorem, roughly speaking, states that if a point lies in a convex hull of a point set, then there is a small witness of that. It was discovered by Carathéodory in 1907 in the context of power series in harmonic analysis.

**Theorem 0.1** (Carathéodory's theorem [Car07]). *Let $P$ be a set of $n$ points in $\mathbb{R}^d$. Then every point in $\operatorname{conv}(P)$ lies in the convex hull of a subset of at most $d+1$ points of $P$.*

Another key property of convexity characterises when a finite family of convex sets has a non-empty intersection. The theorem of Helly from 1930 implies that whenever a family of convex sets does not intersect, there is a small witnessing subfamily which does not intersect either.

**Theorem 0.2** (Helly's theorem [Hel30]). *Let $C_1, ..., C_n$ be convex sets in $\mathbb{R}^d$, $n \geq d+1$. If the intersection of every $d+1$ of these sets is non-empty, then the intersection of all the sets is also non-empty.*

A very important property of convex sets is that they can be linearly separated which is formalised in the following (see e.g. [Mat02, Chapter 1]):

**Theorem 0.3** (Separation theorem). *Let $C, D \subseteq \mathbb{R}^d$ be convex sets with $C \cap D = \emptyset$. Then there is a hyperplane $h$ such that $C$ lies in one closed half-space $h^+$ determined by $h$ and $D$ lies in the opposite closed half-space $h^-$.*

*If $C$ and $D$ are closed and one of them is bounded, then the separation can be made strict, i.e. $C \cap h = \emptyset = D \cap h$. The same is true if $C$ and $D$ are polyhedra.*

A useful result, which simplifies several arguments about intersection patterns of point sets, is the following theorem proved by Radon in 1921.

**Theorem 0.4** (Radon's lemma [Rad21]). *Let $P$ be a set of $n \geq d+2$ points in $\mathbb{R}^d$. Then there are disjoint subsets $A, B \subseteq P$ such that*

$$\operatorname{conv}(A) \cap \operatorname{conv}(B) \neq \emptyset.$$

The pair of sets $A, B$ from the theorem is called a *Radon partition* of $P$ and a point $x \in \operatorname{conv}(A) \cap \operatorname{conv}(B)$ is called a *Radon point.*

Radon's lemma guarantees a partition of any sufficiently large set of points into two subsets whose convex hulls intersect. Furthermore, if a $d+2$ point set is in general position then the Radon partition is known to be unique. A natural question to ask is whether some similar partitioning is possible into three or more subsets. The ultimate answer was given by Tverberg [Tve66] in 1966 and alternative proofs are due to Tverberg [Tve81] and Sarkaria [Sar92].

**Theorem 0.5** (Tverberg's theorem; [Tve66]). *Let $r \in \mathbb{N}$ and $n \geq (r-1)(d+1)+1$. Then for any set $P$ of $n$ points in $\mathbb{R}^d$, there exists $r$ pairwise disjoint subsets $A_1, \ldots, A_r \subseteq P$ such that*

$$\bigcap_{i \in [r]} \operatorname{conv}(A_i) \neq \emptyset$$

Several variants and extensions of this theorem have been researched, including the coloured Tverberg theorem conjectured in the context of $k$-sets in [BFL90] and proved in [ŽV92]. Sharper bounds have been discovered recently in [BMZ09].

**Moment curve.** It is often difficult to come up with upper bound constructions for many questions in higher dimensions. Fortunately, there are several common constructions which happen to work for a range problems. One of these constructions is a set of points on the moment curve and we will use it, almost exclusively, for our upper bounds.

The *moment curve* in $\mathbb{R}^d$ is the curve $\gamma(t) = (t, t^2, \ldots, t^d)$. It intersects any hyperplane $h$ in at most $d$ points. If $h$ and $\gamma$ have $d$ intersections, then all of these are proper, i.e. the curve passes from one side of the hyperplane to the other. On the other hand, any $d$ points define a hyperplane and one can freely choose the intersection points and draw the hyperplane through them. As $t \to \infty$, the points $\gamma(t)$ lie above any fixed non-vertical hyperplane.

Any $n$ parameter values $0 < t_1 < \ldots < t_n$ define a point set

$$C_{n,d} := \{\gamma(t_1), \ldots, \gamma(t_n)\}.$$

The convex hull of such a point set $C_{n,d}$ is called a *cyclic polytope.* Its combinatorial structure, as well as the combinatorial structure of the $k$-facets of $C_{n,d}$, is independent of the particular choice of the parameter values. The polytope $C_{n,d}$ maximises the number of faces of each dimension.

## 0.4 What comes next?

We have given a recap of all the notions that appear throughout this thesis without any explanations. Now we proceed with the three main topics of this thesis:

- *circle containment* and *centre points* (Part I),
- *crossing identities* and *higher dimensional crossings* (Part II), and
- *conflict-free colouring* (Part III).

Each of the parts is self-contained and starts with a chapter containing background information, such as definitions, relevant results and examples. Further chapters then contain our results in the area.

# Part I

# Points, circles and other suspects

# 1 Circle containment: background

Part I of this thesis deals with the circle containment problem which originated in the late 1980s. The question is the following: given a set $P$ of $n$ points in the plane, is there always a pair of points $\{a, b\}$, such that every disc containing $\{a, b\}$ also contains a constant fraction of all the points in $P$? A similar question was asked in higher dimensions as well.

This was initially studied by Neumann-Lara and Urrutia [NLU88] who proved the first bounds on the fraction. Several subsequent papers further investigated this problem and its variants (e.g. higher dimension or only diametral discs) and gave improved bounds.

In Section 1.1, we present an overview of results about the circle containment problem and its history. As a way to tackle this problem, we translate the question to the setting of half-space depth and $r$-centre points. These notions are defined and their relationship to circle containment is explained in Section 1.2 where we also mention some basic structural properties of $r$-centres. The detour through $r$-centre points has the advantage that it has implications in several other settings. In Section 1.3 we formulate the questions which we are interested in or, more precisely, for which our method works: the *ball depth problem* (this is the name we will use for the analogue of circle containment problem in an arbitrary dimension), *half-space depth problem*, *stabbing centre region problem*, *pinning simplices problem* and *intersecting partitions problem.*

We conclude this chapter in Section 1.4 by introducing the tools and facts needed for our proofs. We start with our main tool, the Clarkson-Shor random sampling method, which we illustrate on a classical example. Afterwards we list several auxiliary facts needed for calculations with binomial coefficients.

In Chapter 2, we establish relationships between the existence of a small subset of high half-space depth in a point set, and the other problems of our interest. At that point, we will already know that small point sets of large half-space depth imply small point sets of large ball depth one dimension lower. We will show that if a point set has a large half-space depth, its affine hull intersects an $r$-centre region (where the $r$ is smaller than the depth) and also intersects many simplices spanned by disjoint subsets of vertices. As the last relationship, we will see that high half-space depth of a subset of a point set implies that the subset participates in a partition of the whole point set into many intersecting parts. The above quantifications of large, small and many do, of course, differ.

In Chapter 3, we use the random sampling technique to show the existence of a small subset of high half-space depth in a point set. By the discussion of consequences of high half-space depth from Chapter 2, this implies lower bounds on all the above mentioned problems. Consequently, we also get an improved lower bound for the ball depth problem. Our bounds have a much more favourable dependence on the dimension than the previous bound of [BSSU89] and yield an improvement for $d \geq 4$.

In Chapter 4, we unravel several other properties of the problems of our interest. Namely, we exhibit upper bounds for the problems and discuss algorithmic implications of our lower bound proofs.

## 1.1 Circle containment

In 1988, Neumann-Lara and Urrutia [NLU88] showed that for every set $P$ of $n \geq 2$ points in the plane, there is a pair $\{a, b\} \subseteq P$ of *disc depth* at least $\frac{n-2}{60}$ with respect to $P$, i.e. every disc $D \supseteq \{a, b\}$ contains at least $\frac{n-2}{60}$ points of $P$. The constant $1/60$ was repeatedly improved in a series of subsequent papers by Hayward, Rappaport and Wenger [HRW89], by Bárány, Schmerl, Sidney and Urrutia [BSSU89], and by Hayward [Hay89].

The best lower bound to date was obtained by Edelsbrunner, Hasan, Seidel, and Shen [EHSS89], who proved* that there is always a pair $\{a, b\} \subseteq P$ of disc

*The proof is rather involved and combines an analysis of higher-order Voronoi diagrams with a lower bound of $3\binom{k+1}{2}$ for the number of so-called $(\leq k)$-*sets* of a point set, i.e. of subsets of size at most $k$ that can be separated by a line, $0 \leq k < n/2$. The latter bound is tight for $k \leq n/3$ which is the range in which it is needed for the circle containment problem.

depth at least

$$\frac{n-3}{2} - \sqrt{\frac{n^2-4n+3}{12}} \approx \frac{1}{4.73}n$$

with respect to $P$. A simplified argument was found recently by Ramos and Viaña [RV09].

On the other hand Hayward et al. [HRW89] gave a construction of point sets for which no pair of points from the set has disc depth more than $\lceil n/4 \rceil - 1$. Thus, if we denote by $\mathrm{b}_2(n)$ the largest number such that for every set $P$ of $n$ points in the plane, there is a pair $\{a, b\} \subseteq P$ of disc depth at least $\mathrm{b}_2(n)$, then $n/4.73 \approx (1/2 - 1/\sqrt{12})n + o(n) \leq b_2(n) \leq n/4 + O(1)$. The upper bound is conjectured to be tight [HRW89].

Hayward et al. [HRW89] also considered a variant of the problem in which the point set is required to be in convex position. In this case, they proved that there is always a pair of disc depth at least $\lceil n/3 \rceil + 1$, and that this is tight in the worst case.

Akiyama et al. [AIUU96] considered another variant of the problem where only diametral discs are taken into account (i.e. for every pair of points, one only looks at the diametral disc of these two points, that is, the disc for which the segment connecting the points is a diameter). In that case, they proved that there is always a pair of points whose diametral disc contains $\lceil n/3 \rceil + 1$ points and that this is tight and attained in convex position.

Bárány et al. [BSSU89] considered a natural extension of the original problem to higher dimensions (already raised in [NLU88]). In dimension $d \geq 3$, pairs of points are no longer enough. In fact, a construction based on the moment curve in $\mathbb{R}^{d+1}$ shows that there are $n$-point sets in $\mathbb{R}^d$ for which every subset of size $\lfloor \frac{d+1}{2} \rfloor$ can be separated from the remaining points by a ball, i.e. has ball depth at most $\lfloor \frac{d+1}{2} \rfloor$. Bárány et al. show that this is the threshold: for every $d$, there exists a number $\mathrm{b}_d(n) \in \Theta(n)$ (as a function of $n$) such that for every set $P$ of $n$ points in $\mathbb{R}^d$, there is a subset $X \subseteq P$ of size at most $\lfloor \frac{d+3}{2} \rfloor$ and ball depth at least $\mathrm{b}_d(n)$ with respect to $P$.

By abuse of notation, we denote the largest such number by $\mathrm{b}_d(n)$ as well. We refer to the problem of determining $\mathrm{b}_d(n)$ as the *ball depth problem*. The proof of Bárány et al. yields a lower bound of

$$\mathrm{b}_d(n) \geq \frac{n}{\lceil \frac{d+3}{2} \rceil \binom{d+3}{\lfloor \frac{d+3}{2} \rfloor}} \approx \frac{\sqrt{\pi} n}{\sqrt{\frac{d+3}{2}} \cdot 2^{d+3}},$$

The proof of that bound in [EHSS89] contains a lacuna; correct proofs were later furnished in [LVWW04] and [AFM05] in the context of rectilinear crossing numbers of graphs. See also [AGOR07] for a particularly elegant proof and further improvements.

which is inversely exponential in $d$.

**New results.** In Section 3.2, we give an alternative proof which yields a lower bound that is inversely polynomial (namely, quadratic) in $d$, roughly[1] $\mathrm{b}_d(n) \geq \frac{4n}{(4e+1)d^2}$ in even dimensions and $\mathrm{b}_d(n) \geq \frac{4n}{(3e+1)d^2}$ in odd dimensions. We also use the same idea to prove a variant of the result for pseudo-discs in the plane: in a set of points and pseudo-discs satisfying some technical assumptions, there is always a pseudo-disc containing a $\frac{1}{6e}$-fraction of the points.

## 1.2 Centre points

The way we tackle the ball depth problem as well as the other problems is by a detour via half-space depth, $r$-centre points and $r$-centre regions. Here, we define these three notions and look into the structure of $r$-centre regions.

**Definition 1.1** (Depth). *Let $\mathcal{R}$ be a family of subsets of $\mathbb{R}^d$, which we call* ranges*, and let $P$ be a set of $n$ points in $\mathbb{R}^d$. We define the $\mathcal{R}$-depth of a set $A \subseteq \mathbb{R}^d$ with respect to $P$ ($A$ is not necessarily a subset of $P$) as*

$$\mathrm{depth}_P^{\mathcal{R}}(A) := \min\{|P \cap R| : R \in \mathcal{R}, A \subseteq R\}.$$

The definition is general but in the further text we concentrate on the following families of ranges:

- $\mathcal{B}$, the family of all balls;
- $\mathcal{H}$, the family of all affine half-spaces;
- $\mathcal{H}^-$, the family of all *lower* half-spaces; and
- $\mathcal{H}^+$, the family of all *upper* half-spaces.

We will assume *all point sets that we encounter are in general position.* What exactly that means depends on the ranges in question: in the case of half-spaces, general position means that no $d+1$ or fewer of the points are affinely dependent. In the case of balls, we additionally require that no $d+2$ of the points lie on a common $(d-1)$-dimensional sphere.

We have already encountered ball depth in the description of the circle containment problem. To motivate the definition for half-spaces (from

[1] We only write the leading coefficient in front of $n/d^2$ here. The remaining terms are either sublinear in $n$ or $n \cdot o(1/d^2)$.

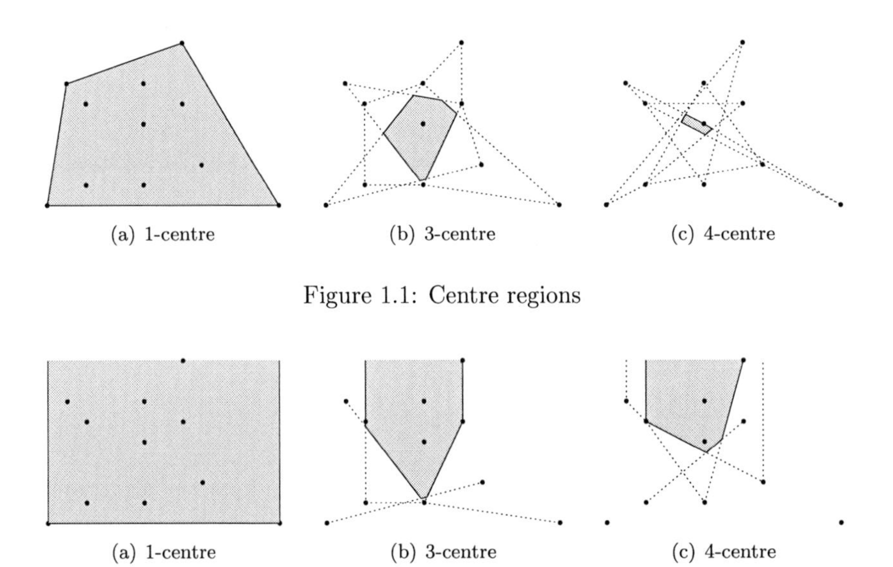

(a) 1-centre (b) 3-centre (c) 4-centre

Figure 1.1: Centre regions

(a) 1-centre (b) 3-centre (c) 4-centre

Figure 1.2: Lower centre regions

the perspective of the circle containment/ball depth problem), we recall the paraboloidal lifting map. Let $U = \{x \in \mathbb{R}^{d+1} : x_{d+1} = x_1^2 + \ldots + x_d^2\}$. The lifting map takes a point $p = (p_1, \ldots, p_d) \in \mathbb{R}^d$ to the point $\hat{p} = (p_1, \ldots, p_d, p_1^2 + \ldots + p_d^2) \in U$. If $B$ is a ball in $\mathbb{R}^d$ then there is a lower half-space$\hat{B}$ in $\mathbb{R}^{d+1}$ (open or closed according to whether $B$ is) such that a point $p \in \mathbb{R}^d$ lies in the interior of $B$, on the boundary of $B$, or outside of $B$, if and only if $\hat{p}$ lies in the interior of $\hat{B}$, on the boundary of $\hat{B}$, or in the upper open half-space complementary to $\hat{B}$, respectively. We refer to [Mat02, Section 5.7] for more details. Thus, if $P$ is a set of points in $\mathbb{R}^d$, then a set $A$ has ball depth $r$ with respect to $P$ if and only if $\hat{A}$ has *lower half-space depth* $r$ with respect to $\hat{P}$.

This allows us, in a sense, to forget about the balls and restrict our attention to the case that all ranges are half-spaces, i.e. we assume that⌝ $\mathcal{R} \subseteq \mathcal{H}$.

**Definition 1.2** (Centre points). *Let $P$ be a set of $n$ points in $\mathbb{R}^d$. For any integer $r \geq 0$ and $\mathcal{R} \subseteq \mathcal{H}$, the* $r$-centre region *of $P$ with respect to $\mathcal{R}$ is the*

⌝One of the reasons is that for other ranges, the upcoming definition of the centre is usually not very meaningful since single points do not have large depth; for instance, in the case $\mathcal{R} = \mathcal{B}$ we have $C_1^{\mathcal{B}}(P) = P$ and $C_r^{\mathcal{B}}(P) = \emptyset$ for $r > 1$.

*set*[§]

$$C_r^{\mathcal{R}}(P) := \{x \in \mathbb{R}^d : \operatorname{depth}_P^{\mathcal{R}}(\{x\}) \geq r\}.$$

*The elements of this set are called* $r$-centre points *with respect to* $\mathcal{R}$.

*When* $\mathcal{R}$ *is* $\mathcal{H}$ *(respectively,* $\mathcal{H}^-$*) we often simply speak of* $r$-centre points *and* regions *(respectively,* lower $r$-centre points *and* regions*), i.e. with no respect.*

Note that $\mathcal{R}' \subseteq \mathcal{R}$ implies $C_r^{\mathcal{R}'}(P) \supseteq C_r^{\mathcal{R}}(P)$ for all $r$ and $P$. An $\frac{n}{d+1}$-centre point with respect to $\mathcal{H}$ is often just called a *centre point*. This can be seen as a higher dimensional generalisation of a median. Every set $P$ of points has a centre point (but this centre point usually does not lie in $P$). We will sketch the proof shortly.

Historically, centre points were studied first, and only later, when it turned out that the exact centre points are difficult to compute, people looked into $r$-centre points which have then been studied mainly from the algorithmic perspective.

A randomised algorithm of Chan [Cha04] finds a centre point of $P$ in expected time $O(n^{d-1})$ which is polynomial in $n$ if the dimension $d$ is fixed. However, if both $n$ and $d$ are part of the input, the problem of testing whether a given point is a centre point is known to be coNP-complete [Ten91]. With the weaker $r$-centre condition, it is possible to find $\Omega(\frac{n}{d^2})$-centres in polynomial time. A randomised algorithm of Clarkson et al. [CEM+93, CEM+96] outputs in time $O(d^2(d \log n)^{\log d+2})$ a point which is an $\Omega(n/d^2)$-centre with arbitrarily small constant probability of error[¶]. A more sophisticated variant of the algorithm presented in the same paper finds an $\Omega(n/d^2)$-centre in time $d^{O(\log d)}$ which is independent of $n$, with arbitrarily small constant probability of error. The last variant thereof finds an $\Omega(n/d^2)$-centre with high probability in polynomial time $O(nd^3(d+\log\log n))$. A fully deterministic sub-exponential algorithm for finding an $\Omega(n/d^2)$-centre in time $n^{O(\log d)}$ has been discovered recently by Miller and Sheehy [MS09, MS10].

Lower $r$-centre regions have apparently not been systematically studied yet but many of their properties are, not surprisingly, similar to those of $r$-centre regions. For further investigation, we will need that (i) $C_r^{\mathcal{H}}(P)$ and $C_r^{\mathcal{H}^-}(P)$ are convex polyhedra and (ii) how big $r$ has to be for $C_r^{\mathcal{H}^-}(P)$ to exist. We will spend the rest of this section discussing these topics.

[§]In the literature one can often find a notion of an $\alpha$-centre point which corresponds to an $\alpha n$-centre point with respect to $\mathcal{H}$ in our definition.

[¶]The algorithm does not know whether its answer is correct or not and this is the case for the other two algorithms as well. The error probability of the algorithm has an influence on the constant in the running time bound.

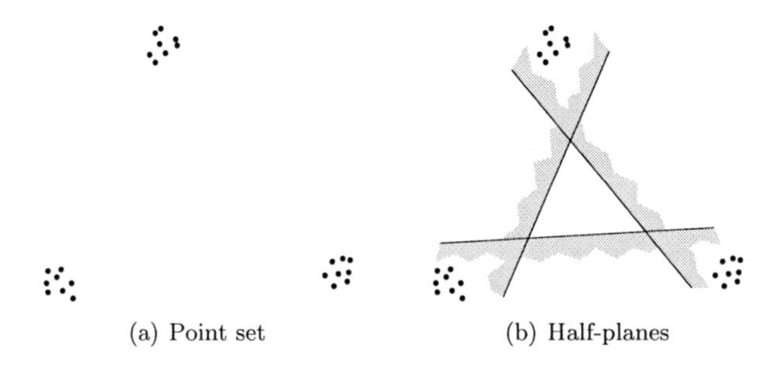

(a) Point set (b) Half-planes

Figure 1.3: Point set with no $(n/3+1)$-centre point

Denote by $\mathcal{H}_{\leq i}(P)$ (respectively, $\mathcal{H}^+_{\leq i}(P)$) the set of all closed half-spaces (respectively, closed upper half-spaces) that *miss*, i.e. whose complementary open half-space contains, at most $i$ points of $P$. It is easy to see that:

**Observation 1.3.** *For every set $P$ of $n$ points in $\mathbb{R}^d$ and $r \leq n/2$,*

$$C_r^{\mathcal{H}}(P) = \bigcap \mathcal{H}_{\leq r-1}(P) \quad \textit{and} \quad C_r^{\mathcal{H}^-}(P) = \bigcap \mathcal{H}^+_{\leq r-1}(P).$$

*Proof.* If $x \notin H \in \mathcal{H}_{\leq r-1}(P)$ then $x \notin C_r^{\mathcal{H}}(P)$ since the complement of $H$ contains a closed half-space containing $x$ and at most $r-1$ points from $P$. On the other hand, if $x \notin C_r^{\mathcal{H}}(P)$ then there is a half-space containing $x$ and at most $r-1$ points of $P$. This half-space can be slightly perturbed and translated such that it has $x$ on its boundary and still contains at most $r-1$ points of $P$, all in the interior. Its closed complementary half-space $H$ contains $x$ and misses at most $r-1$ points of $P$ and consequently, $x \notin H \in \mathcal{H}_{\leq r-1}(P)$. $\square$

If $(d+1)(r-1) < n$, i.e. $r \leq \lceil n/(d+1) \rceil$, then the intersection of any $d+1$ of these half-spaces is non-empty, and Helly's theorem implies $C_r^{\mathcal{H}}(P) \neq \emptyset$. In general this is best possible, i.e. there are examples of $n$-point sets with $C_r^{\mathcal{H}}(P) = \emptyset$ for $r > \lceil \frac{n}{d+1} \rceil$ (see Figure 1.3).

Agarwal et al. [ASW08, Lemma 2.1] showed that in the representation of $C_r^{\mathcal{H}}(P)$ as an intersection of closed half-spaces, it is sufficient to take only those half-spaces $H$ that miss *exactly* $r-1$ points of $P$ and whose bounding hyperplane $\partial H$ is spanned by $d$ points of $P$.

To be precise, define $\overline{\mathcal{H}}_i(P)$ to be the set of all closed half-spaces with bounding hyperplane spanned by $d$ points of $P$ which miss exactly $i$ points of $P$ and let $\overline{\mathcal{H}}_{\leq i}(P) := \cup_{j=0}^{i} \overline{\mathcal{H}}_j(P)$.

**Lemma 1.4** ([ASW08]). *Let $P$ be a set of $n$ points in $\mathbb{R}^d$ in general position. Then*

$$C_r^{\mathcal{H}}(P) = \bigcap \overline{\mathcal{H}}_{\leq r-1}(P) = \bigcap \overline{\mathcal{H}}_{r-1}(P).$$

*Consequently, $C_r^{\mathcal{H}}(P)$ is a convex polytope.*

This settles the first question for the $r$-centre regions: they are polytopes. We now turn to lower half-spaces where a similar description can be found. In this case, the centre is still a convex polyhedron, but it is unbounded (if it is non-empty). We state this more precisely in the next lemma but first, let us introduce the necessary notation.

By $\overline{\mathcal{H}}^+_{\leq i}(P)$ we denote the subset of upper half-spaces $\overline{\mathcal{H}}_{\leq i}(P) \cap \mathcal{H}^+$. We identify $\mathbb{R}^{d-1}$ with the hyperplane $\{x_d = 0\}$ and denote by $\pi$ the orthogonal projection onto that hyperplane. Analogously to the proof of [ASW08, Lemma 2.1], one can show the following:

**Lemma 1.5.** *Let $P$ be a set of $n$ points in $\mathbb{R}^d$ in general position. Then*‖

$$C_r^{\mathcal{H}^-}(P) = \bigcap \mathcal{H}^+_{\leq r-1}(P) = \left(\bigcap \overline{\mathcal{H}}^+_{\leq r-1}(P)\right) \cap \left(C_r^{\mathcal{H}}(\pi(P)) \times \mathbb{R}\right).$$

*Consequently, $C_r^{\mathcal{H}^-}(P)$ is a convex polyhedron.*

In other words, one can take all the half-spaces that might possibly be defining the lower part of the usual $r$-centre region and cut them off by the $r$-centre region of the projection extended in the vertical direction.

The first equality is exactly the description of (lower) $r$-centre regions from Observation 1.3. The second equality needs a proof:

*Proof.* We start with "$\subseteq$". The inclusion $\bigcap \mathcal{H}^+_{\leq r-1}(P) \subseteq \bigcap \overline{\mathcal{H}}^+_{\leq r-1}(P)$ follows from the fact that the family on the right is a subfamily of the family on the left. To see $\bigcap \mathcal{H}^+_{\leq r-1}(P) \subseteq C_r^{\mathcal{H}}(\pi(P)) \times \mathbb{R}$, consider a point $x \notin C_r^{\mathcal{H}}(P) \times \mathbb{R}$. Then there is a vertical half-space $H'$ containing $x$ and at most $r-1$ points of $P$ with no points on its boundary. Then a small perturbation of $H'$ yields a lower half-space $H \ni x$ containing at most $r-1$ points of $P$ in its interior and none on its boundary. Its complementary half-space misses at most $r-1$ points of $P$ and does not contain $x$. Thus, $x \notin \bigcap \mathcal{H}^+_{\leq r-1}(P)$.

For the inclusion "$\supseteq$" consider any half-space $H \in \mathcal{H}^+_{\leq r-1}(P)$. Lemma 1.4 implies $C_r^{\mathcal{H}}(\pi(P)) = \bigcap \overline{\mathcal{H}}^+_{\leq r-1}(\pi(P))$. We will find a set of half-spaces whose

‖To prove that the lower $r$-centre region is a convex polyhedron, one only needs an analogue of the first equality in Lemma 1.4. An additional step in the proof would get an analogous result with $\overline{\mathcal{H}}_i(P)$ instead of $\overline{\mathcal{H}}_{\leq i}(P)$ in Lemma 1.5.

intersection lies in $H$, each of which misses at most $r-1$ points of $P$ and is either an upper half-space spanned by $d$ points of $P$ or a vertical half-space spanned by $d-1$ points of $P$: Consider $S := H \cap P$ and consider the set $\mathcal{H}_S$ of all the facet-supporting upper half-spaces of $\operatorname{conv}(S)$ together with all the facet supporting half-spaces of $\operatorname{conv}(\pi(S))$ extended to the vertical direction. Their intersection is contained in $H$ which proves the inclusion. □

**Corollary 1.6.** *If $r \leq \lceil n/d \rceil$ then $C_r^{\mathcal{H}^-}(P) \neq \emptyset$, and this is tight.*

*Proof.* If $r \leq \lceil n/d \rceil$, consider any point $(x_1, \ldots, x_{d-1}) \in C_r^{\mathcal{H}}(\pi(P))$ (which we already know to be non-empty from earlier). By choosing a sufficiently large last coordinate $x_d$, we can lift it to a point $x = (x_1, \ldots, x_d)$ that lies above all hyperplanes spanned by $P$, and then $x \in C_r^{\mathcal{H}^-}(P)$. On the other hand, if $r > \lceil n/d \rceil$, we can first choose an $n$-point set $P' \subseteq \mathbb{R}^{d-1}$ with $C_r^{\mathcal{H}}(P') = \emptyset$. Then, by the preceding lemma, any lifting of $P'$ to $\mathbb{R}^d$, i.e. any $P \subseteq \mathbb{R}^d$ with $\pi(P) = P'$ will have $C_r^{\mathcal{H}^-}(P) = \emptyset$. □

## 1.3 Questions

With our newly acquired fluency in the language of depth and $r$-centre regions we can now state the questions we will study. Let us list them first and afterwards, we relate them to each other. An overview of the listed problems and their other variants is included in Figure 1.4 on the next page.

**Ball depth.** What is the largest number $\mathrm{b}_d(n)$ such that every set $P$ of $n$ points in $\mathbb{R}^d$ in general position contains a subset $X \subseteq P, |X| \leq \lfloor \frac{d+1}{2} \rfloor + 1$ with $\operatorname{depth}_P^{\mathcal{B}}(X) \geq \mathrm{b}_d(n)$?

**Lower half-space depth.** What is the largest number $\mathrm{lh}_d(n)$ such that every set $P$ of $n$ points in $\mathbb{R}^d$ in general position contains a subset $X \subseteq P, |X| \leq \lfloor \frac{d}{2} \rfloor + 1$ with $\operatorname{depth}_P^{\mathcal{H}^-}(X) \geq \mathrm{lh}_d(n)$?

**Half-space depth.** What is the largest number $\mathrm{h}_d(n)$ such that every set $P$ of $n$ points in $\mathbb{R}^d$ in general position contains a subset $X \subseteq P, |X| \leq \lfloor \frac{d}{2} \rfloor + 1$ with $\operatorname{depth}_P^{\mathcal{H}}(X) \geq \mathrm{h}_d(n)$?

**Stabbing centre region.** What is the largest number $\mathrm{r}_d(n)$ such that every set $P$ of $n$ points in $\mathbb{R}^d$ in general position contains a subset $X \subseteq P, |X| \leq \lfloor \frac{d}{2} \rfloor + 1$ with $\operatorname{conv}(X) \cap C_{\mathrm{r}_d(n)}^{\mathcal{H}}(P) \neq \emptyset$?

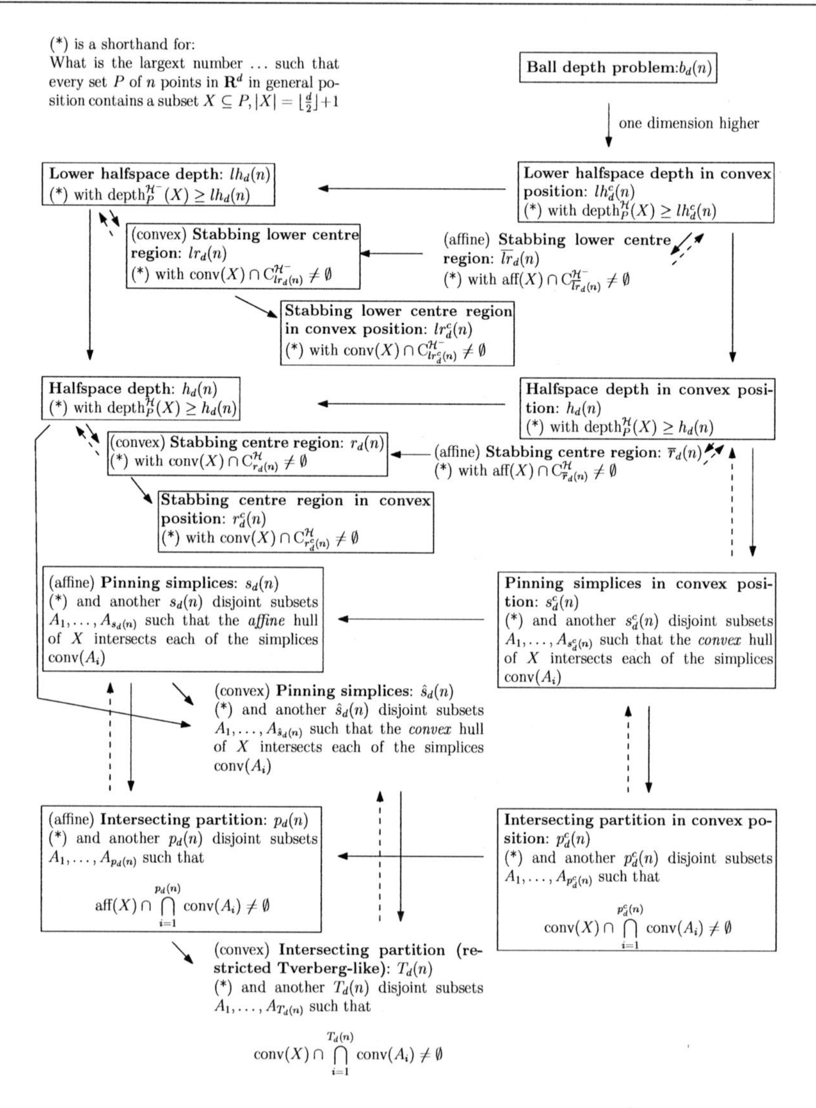


Figure 1.4: Overview of the questions. The ones in rectangles will be studied. Arrows indicate relationships: full means source $\geq$ target; dashed indicate a weaker relation (inequality between multiples).

**Stabbing lower centre region.** What is the largest number $\mathrm{lr}_d(n)$ such that every set $P$ of $n$ points in $\mathbb{R}^d$ in general position contains a subset $X \subseteq P, |X| \leq \lfloor \frac{d}{2} \rfloor + 1$ with $\mathrm{conv}(X) \cap C^{\mathcal{H}^-}_{\mathrm{lr}_d(n)}(P) \neq \emptyset$?

**Pinning simplices.** What is the largest number $\mathrm{s}_d(n)$ such that every set $P$ of $n$ points in $\mathbb{R}^d$ in general position contains a subset $X \subseteq P, |X| \leq \lfloor \frac{d}{2} \rfloor + 1$ and $\mathrm{s}_d(n)$ other pairwise disjoint subsets $A_1, \ldots, A_{\mathrm{s}_d(n)} \subseteq P \setminus X$ such that the affine hull of $X$ intersects each of the simplices $\mathrm{conv}(A_i)$?

**Intersecting partition.** What is the largest number $\mathrm{p}_d(n)$ such that every set $P$ of $n$ points in $\mathbb{R}^d$ in general position contains $\mathrm{p}_d(n) + 1$ pairwise disjoint subsets $X, A_1 \ldots, A_{\mathrm{p}_d(n)} \subseteq P$ such that $|X| \leq \lfloor \frac{d}{2} \rfloor + 1$ and

$$\mathrm{aff}(X) \cap \bigcap_{i=1}^{\mathrm{p}_d(n)} \mathrm{conv}(A_i) \neq \emptyset?$$

One might wonder, why are the sizes of the sets $X$ chosen the way they are. The reason is simple: smaller values would result in trivial questions, since a point set on a moment curve would attain value 0 for all of them. This was proved in [BSSU89] for the ball depth problem. For all the other problems it is a consequence of the fact that any $\lfloor \frac{d}{2} \rfloor$ points in a set $P$ of $n$ points on the moment curve form a facet of $\mathrm{conv}(P)$. Nevertheless, one could also ask the same for larger sets $X$. This is a perfectly valid question! Although we do not pursue this direction explicitly, our methods do have implications (and yield an improvement) for any $X$ larger than prescribed and it is not difficult to trace the key lemmata (which are usually phrased more generally) and calculate the bounds for these situations.

We can also consider each of these problems only for point sets $P$ in convex position. Then one can also replace the affine spans by convex hulls in the problem definitions. To express that we mean this variant of a given problem, we will add the phrase "in convex position" after the name of the problem (e.g. lower half-space depth in convex position) and add a superscript $^c$ to the function (e.g. $\mathrm{lh}^c_d(n)$).

Notice that the intersecting partition problem in convex position is very similar to Tverberg's theorem: the only difference is that we only allow point sets in convex position and in return we ask for one part to be small.

**Relationships between the problems.** Clearly, we have $\mathrm{b}^c_d(n) \geq \mathrm{b}_d(n)$, $\mathrm{h}^c_d(n) \geq \mathrm{h}_d(n), \mathrm{lh}^c_d(n) \geq \mathrm{lh}_d(n), \mathrm{r}^c_d(n) \geq \mathrm{r}_d(n), \mathrm{lr}^c_d(n) \geq \mathrm{lr}_d(n), \mathrm{s}^c_d(n) \geq \mathrm{s}_d(n)$

and $\mathrm{p}_d^c(n) \geq \mathrm{p}_d(n)$. Our earlier discussion about the lifting map implies the inequality $\mathrm{b}_d(n) \geq \mathrm{lh}_{d+1}^c(n)$ since every point set in $\mathbb{R}^d$ can be mapped to a point set in $\mathbb{R}^{d+1}$ in convex position where, vaguely speaking, balls are mapped to lower half-spaces. Furthermore, we have $\mathrm{lh}_d(n) \geq \mathrm{h}_d(n), \mathrm{lh}_d^c(n) \geq \mathrm{h}_d^c(n), \mathrm{lr}_d(n) \geq \mathrm{r}_d(n)$ and $\mathrm{lr}_d^c(n) \geq \mathrm{r}_d^c(n)$ since $\mathcal{H}^- \subseteq \mathcal{H}$. Note that if a convex set intersects an $r$-centre region then it must also have half-space depth at least $r$ and hence, $\mathrm{h}_d(n) \geq \mathrm{r}_d(n)$ and $\mathrm{h}_d^c(n) \geq \mathrm{r}_d^c(n)$.

From the other end, if a convex set intersects an intersection of convex sets then it also intersects each of them separately and we have $\mathrm{s}_d(n) \geq \mathrm{p}_d(n)$ and $\mathrm{s}_d^c(n) \geq \mathrm{p}_d^c(n)$. The next step only works in the convex setting: if a convex body intersects a convex polytope then also every half-space containing the convex body contains at least one vertex of the polytope. This implies $\mathrm{h}_d^c(n) \geq \mathrm{s}_d^c(n)$.

Consequently, lower bounds on each of $\mathrm{p}_d^c(n), \mathrm{s}_d^c(n)$ and both convex and non-convex variants of $\mathrm{r}_d(n), \mathrm{r}_d^c(n), \mathrm{lr}_d(n), \mathrm{lr}_d^c(n), \mathrm{lh}_d(n), \mathrm{lh}_d^c(n), \mathrm{h}_d(n)$, as well as $\mathrm{h}_d^c(n)$, are also lower bounds on the ball depth problem one dimension lower, i.e. $\mathrm{b}_{d-1}(n)$. Hence, studying each of these problems might be interesting for improving the lower bounds on the ball depth problem. We will, however, try to go in the opposite direction, i.e. to find out what we can say about the other problems if we know $\mathrm{h}_d(n)$.

## 1.4 Tools and methods

In the following chapters, we use the random sampling method of Clarkson and Shor, which will be discussed shortly, and several auxiliary identities involving binomial coefficients which appear at the end of this section.

### 1.4.1 Random sampling

The heart of our lower bound proof is the random sampling technique of Clarkson and Shor [CS89]. Here, we describe the idea of the technique and demonstrate it on a short example. We follow the abstract framework of Sharir [Sha01, Sha03] throughout.

Consider a set $S$ of $n$ *objects* and a set $C \subseteq \binom{S}{d}$ of *configurations*, each defined by $d$ objects of $S$ (for some constant integer parameter $d$). We are given a set of *conflicts* $X \subseteq S \times C$ between the objects and the configurations – where we assume that a configuration $c \in C$ is not in conflict with the objects defining it. The *weight* $w_S(c)$ of a configuration $c \in C$ is the number of objects $a \in S$ in conflict with $c$, i.e. such that $(a, c) \in X$.

For a concrete example, consider a set $S$ of $n$ lines in general position in the plane. As configurations, we consider vertices of the arrangement of the lines, i.e. their intersections. A vertex is in conflict with a line if it lies strictly above it. The vertices of weight $k$ are then the upper $k$-level vertices of the arrangement.

For a subset $C' \subseteq C$ of configurations and a subset $X \subseteq S$ of objects, denote by** $C'_k(X)$ the set of all configurations of weight $k$ (the set $X$ might be a proper subset of $S$ and only the conflicts with the objects in $X$ are considered here) and $C'_{\leq k}(X)$ the set of those of weight at most $k$.

By random sampling, one can turn upper bounds on $|C'_0|$ into upper bounds on $|C'_{\leq k}|$ by the following method:

Consider a random sample $S' \subseteq S$ by keeping each object $a \in S$ independently with probability $p$ (a parameter, to be determined later). This induces a random subset $C' \subseteq C$ of configurations whose defining objects are in $S'$. Denote by $A_c$ the event that a configuration $c \in C$ remains in the random sample and has weight 0 there, i.e. the event $c \in C'_0(S')$. Then we claim:

$$\mathbf{E}\,|C'_0(S')| = \sum_{c\in C} \mathbf{Pr}(A_c) = \sum_{c\in C} p^d(1-p)^{w_S(c)}.$$

The second equality follows from the fact that the $d$ objects defining $c$ have to survive and all its $w_S(c)$ conflicting objects have to disappear during the random sampling. Furthermore, we have:

$$\sum_{c\in C} p^d(1-p)^{w_S(c)} \geq \sum_{\substack{c\in C\\ w_S(c)\leq k}} p^d(1-p)^{w_S(c)} \geq \sum_{\substack{c\in C\\ w_S(c)\leq k}} p^d(1-p)^k = |C_{\leq k}(S)|p^d(1-p)^k$$

Consequently,

$$|C_{\leq k}(S)| \leq \frac{\mathbf{E}\,|C'_0(S')|}{p^d(1-p)^k}. \tag{1.2}$$

Finding an optimal value of $0 < p < 1$ then yields some particular upper bound.

Let us finish the example with line arrangements. There are at most $n-1$ vertices on the 0-level in an arrangement of $n$ lines since the lines on the 0-level (a.k.a. the lower envelope) appear in the order of decreasing slopes and therefore, no line appears twice. This implies

$$|C_{\leq k}(S)| \leq \frac{pn}{p^2(1-p)^k}.$$

**Here $C'$ really is the name of the set $C'$ of configurations. That is, if the set of configuration was called $C''$ we would write $C''_k(X)$.

Substituting $p := \frac{1}{k+1}$ yields

$$|C_{\leq k}(S)| \leq \frac{n(k+1)}{(1-\frac{1}{k+1})^k} \leq n(k+1)(1+\tfrac{1}{k})^k \leq en(k+1).$$

It is common in many situations that the chosen sample (e.g. for $k$ linear in $n$ in the the above example) has expected constant size. This often has favourable consequences for approximation algorithms.

### 1.4.2 Auxiliary facts

We need to work quite a lot with binomial coefficients to be able to obtain reasonable bounds from the random sampling argument for the half-space depth problem and we will use the following facts:

**Fact 1.7** (Pascal triangle). *Let $a, b \in \mathbb{Z}_0^+, a < b$, then*

$$\binom{a}{b} + \binom{a}{b+1} = \binom{a+1}{b+1}.$$

**Fact 1.8.** *Let $a, b \in \mathbb{Z}_0^+$. Then*

$$\sum_{i=0}^{b} \binom{a+i}{a} = \binom{a+b+1}{a+1}.$$

**Fact 1.9.** *Let $N$ be a binomially distributed random variable $N \sim \mathrm{B}(n,p)$ and let $s \in \mathbb{N}$. Then*

$$\mathbf{E}\left[\binom{N}{s}\right] = p^s \binom{n}{s}.$$

The first one is a well known identity whereas the other two may be slightly less standard and therefore, we include their proofs in Section A.2 of the Appendix.

# 2 The geometry: depth and centres

In this chapter, we establish connections between the half-space depth problem and the problems of stabbing centre region, pinning simplices and intersecting partition.

In Section 2.1, we show a relationship between the half-space depth problem and stabbing centre regions problem. We already know that $\mathrm{h}_d(n) \geq \mathrm{r}_d(n)$. Here we find a bound in the other direction: $(\lfloor \frac{d}{2} \rfloor + 1)\, \mathrm{r}_d(n) \geq \mathrm{h}_d(n)$ in Corollary 2.3. As a key step along the way, we prove that for a set $X$ whose convex hull is disjoint from an $r$-centre region, one can find $|X|$ facet-supporting half-spaces of the centre region and a half-space containing $X$ which lies in the union of their complements.

In Section 2.2, we show a link between the half-space depth problem and the pinning simplices problem. From the previous chapter, we know that $\mathrm{h}_d(n) \geq \mathrm{s}_d(n)$ but here we pursue the opposite direction and prove the inequality $\lceil \frac{d}{2} \rceil\, \mathrm{s}_d(n) \geq \mathrm{h}_d(n)$. The proof is based on an easy greedy argument.

In Section 2.3, we show the last relation of this chapter – between the pinning simplices problem and the intersecting partition problem. We know that $\mathrm{s}_d(n) \geq \mathrm{p}_d(n)$ but nothing in the opposite direction. Here, we prove $(\lceil \frac{d}{2} \rceil + 1)\, \mathrm{p}_d(n) \geq \mathrm{s}_d(n)$ which finishes our chain. Together with the relationship between half-space depth and pinning simplices, this implies the inequality $(\lceil \frac{d}{2} \rceil + 1)\lceil \frac{d}{2} \rceil \, \mathrm{p}_d(n) \geq \mathrm{h}_d(n)$.

This is joint work with Shakhar Smorodinsky and Uli Wagner [SSW08].

## 2.1 Stabbing centre regions

We start with the relationship between the half-space depth problem and the stabbing centre regions problem. First, let us recall the problem definitions:

**Half-space depth.** What is the largest number $\mathrm{h}_d(n)$ such that every set $P$ of $n$ points in $\mathbb{R}^d$ in general position contains a subset $X \subseteq P, |X| \leq \lfloor \frac{d}{2} \rfloor + 1$ with $\mathrm{depth}_P^{\mathcal{H}}(X) \geq \mathrm{h}_d(n)$?

**Stabbing centre region.** What is the largest number $\mathrm{r}_d(n)$ such that every set $P$ of $n$ points in $\mathbb{R}^d$ in general position contains a subset $X \subseteq P, |X| \leq \lfloor \frac{d}{2} \rfloor + 1$ with $\mathrm{conv}(X) \cap C^{\mathcal{H}}_{\mathrm{r}_d(n)}(P) \neq \emptyset$?

We have observed earlier that any point set whose convex hull intersects an $r$-centre region has half-space depth at least $r$ and consequently,

$$\mathrm{h}_d(n) \geq \mathrm{r}_d(n) \quad \text{and} \quad \mathrm{h}^c_d(n) \geq \mathrm{r}^c_d(n).$$

Can we say something about the converse? As it turns out, the values $\mathrm{h}_d(n)$ and $\mathrm{r}_d(n)$ are not too far apart:

$$(\lfloor \tfrac{d}{2} \rfloor + 1)\, \mathrm{r}_d(n) \geq \mathrm{h}_d(n) \quad \text{and} \quad (\lfloor \tfrac{d}{2} \rfloor + 1)\, \mathrm{r}^c_d(n) \geq \mathrm{h}^c_d(n).$$

In this section we prove these inequalities. First, let us sketch the big picture of the proof. We need to prove that the convex hull of a set $X$ of half-space depth at least $(r-1)\cdot|X|+1$ intersects the $r$-centre region. We prove the contrapositive, that is, if $\mathrm{conv}(X)$ does not intersect the $r$-centre region then there is a half-space containing $X$ and at most $(r-1)\cdot|X|$ points. To do this, we observe that $X$ lies in the union $U$ of at most $|X|$ complements of facet-supporting half-spaces of the $r$-centre region. Each such complementary half-space can contain at most $r-1$ points and hence, $U$ contains at most $|X|(r-1)$ points in total. A half-space which lies completely in $U$, and consequently contains $X$ and at most $|X|(r-1)$ points, is what we wanted.

First, we prove a general lemma which allows us to do the main step, i.e. find a half-space containing $X$ which lies in a union of only $\dim(\mathrm{aff}(X))+1 \leq |X|$ complements of facet-supporting half-spaces of the $r$-centre region. The statement of the lemma is illustrated in Figure 2.1 on the facing page.

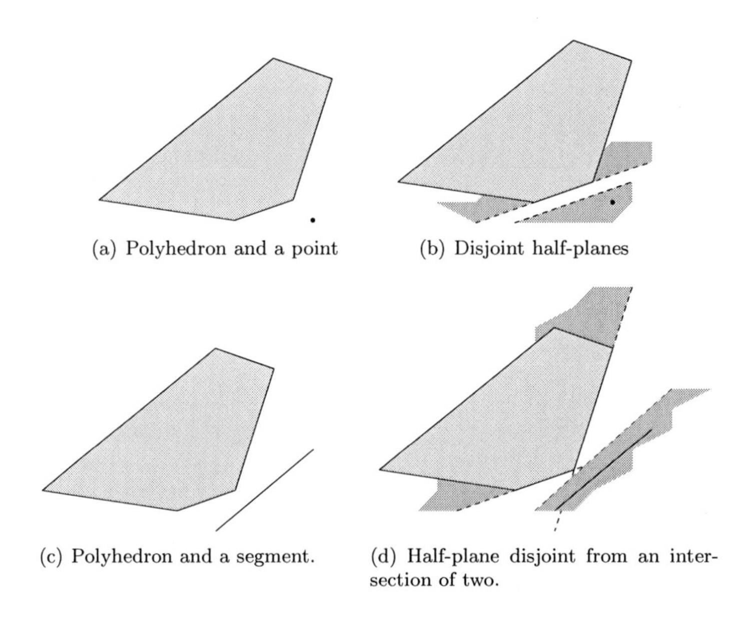

(a) Polyhedron and a point

(b) Disjoint half-planes

(c) Polyhedron and a segment.

(d) Half-plane disjoint from an intersection of two.

Figure 2.1: Illustration of Lemma 2.1.

**Lemma 2.1.** *Let $Q \subseteq \mathbb{R}^d$ be a full-dimensional convex polyhedron and let $K \subseteq \mathbb{R}^d$ be a $k$-dimensional convex body that is disjoint from $Q$ and is either compact or polyhedral. Then there exist $m = \min\{d, k+1\}$ facet-supporting closed half-spaces $H_1, \ldots, H_m$ of $Q$ and a half-space $H$ containing $K$ such that $H \cap H_1 \cap \ldots \cap H_m = \emptyset$.*

*Proof.* Since $K$ and $Q$ are disjoint, both are closed, and $K$ is compact or polyhedral, we can strictly separate them. That is, there is a hyperplane $h$ such that $Q$ is contained in one of the corresponding open half-spaces, which we denote by $h^+$, and $K$ is contained in the opposite open half-space, $h^-$. Moreover, since $Q$ and the hyperplane $h$ are polyhedral and disjoint, they have positive distance from each other. The points $a \in Q$ for which $\text{dist}(a, h) = \text{dist}(Q, h)$ form a face $F$ of $Q$ of some dimension $d - r$, $1 \leq r \leq d$, and a suitable translate of $h$ is a hyperplane supporting $Q$ in $F$. Let $h_1, \ldots, h_t$, be the facet-supporting hyperplanes of $Q$ defining $F$ (i.e. those which have a non-empty intersection with $F$; this intersection has to be a face of $F$ since $Q$ is a convex polyhedron), and let $H_1, \ldots, H_t$ be the corresponding closed half-spaces containing $Q$ (that means, $F \subseteq Q \subseteq H_1 \cap \ldots \cap H_t$). Then we claim

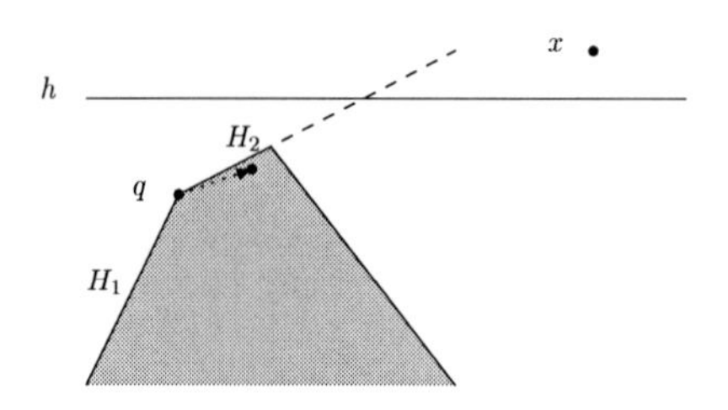


Figure 2.2: Illustration of the distance argument.

that $h^- \cap H_1 \cap \ldots \cap H_t = \emptyset$.

To see this, assume that there exists some $x \in h^- \cap H_1 \cap \ldots \cap H_t$, and let $q \in F$. Since $x, q \in H_1 \cap \ldots \cap H_t$, any convex combination of $x$ and $q$ lies in $H_1 \cap \ldots \cap H_t$. Moreover, if $H$ is a facet-supporting half-space of the polyhedron $Q$ other than $H_1, \ldots, H_t$, then $q$ lies in the interior of $H$. Thus, for sufficiently small $\varepsilon > 0$, the point $p := (1 - \varepsilon)q + \varepsilon x$ still lies in $Q$ which is a contradiction, since it has smaller distance to $h$ than $q$.

Now, we only need to reduce the number of half-spaces. Let $A$ denote the affine hull of $K$. Then $\emptyset = A \cap h^- \cap H_1 \cap \ldots \cap H_t = (A \cap h^-) \cap (A \cap H_1) \cap \ldots \cap (A \cap H_t)$. Thus, we have $t+1$ convex sets in the $k$-dimensional space $A$ that have an empty intersection. By Helly's theorem, there is a subfamily of at most $k+1$ of these sets that have an empty intersection (note that if $A$ is full-dimensional then $h^- = A \cap h^-$ has to be one of the $k+1$ sets since the remaining ones do intersect). Hence, by adding $h^- \cap A$ to this subfamily and by relabelling the $H_i$, if either of these is necessary, we may assume that $A \cap h^- \cap H_1 \cap \ldots \cap H_m = \emptyset$, where $m = \min\{d, k+1\}$. Since $K \subseteq A \cap h^-$, we conclude that $K$ is disjoint from the polyhedron $H_1 \cap \ldots \cap H_m$. Therefore, we can separate them, i.e. find a half-space $H$ containing $K$ such that $H \cap H_1 \cap \ldots \cap H_m = \emptyset$ as desired. □

One can now apply the lemma to $Q := C_r^{\mathcal{R}}$ for $\mathcal{R} = \mathcal{H}$ or $\mathcal{R} = \mathcal{H}^-$ and an arbitrary compact or polyhedral convex body $K$. Here, we need Lemma 1.4 and Lemma 1.5 which guarantee that $C_r^{\mathcal{H}}$ is a convex polytope and $C_r^{\mathcal{H}^-}$ is a convex polyhedron. Every facet-supporting half-space of $C_r^{\mathcal{H}}$ and $C_r^{\mathcal{H}^-}$ misses at most $r-1$ points by the description of the polyhedron as an intersection of finitely many half-spaces missing at most $r-1$ points in Lemma 1.4 and Lemma 1.5. Consequently, each of the half-spaces complementary to $H_1, \ldots, H_m$ contains at most $r-1$ points. Hence, their union contains at most $m(r-1)$ points and therefore, $|H \cap P| \leq m(r-1)$.

**Corollary 2.2.** *If $\mathcal{R} = \mathcal{H}$ or $\mathcal{R} = \mathcal{H}^-$, and if $K$ is a convex body of dimension*

*$k$ which is compact or polyhedral and $K \cap C_r^{\mathcal{R}}(P) = \emptyset$, then* $\operatorname{depth}_P^{\mathcal{R}}(K) \leq m(r-1)$ *where* $m = \min\{d, k+1\}$.

Let us return to the setting of stabbing centre regions. If $r > \mathrm{r}_d(n)$ then there must be a set $P \subseteq \mathbb{R}^d, |P| = n$ such that each of its subsets $X \subseteq P, |X| = \lfloor \frac{d}{2} \rfloor + 1$ has $\operatorname{conv}(X) \cap C_r^{\mathcal{H}}(P) = \emptyset$. The corollary implies that

$$\operatorname{depth}_P^{\mathcal{H}}(X) = \operatorname{depth}_P^{\mathcal{H}}(\operatorname{conv}(X)) \leq (\lfloor \tfrac{d}{2} \rfloor + 1)(r-1),$$

for each such $X$. Consequently, all subsets $X \subseteq P$ of cardinality at most $\lfloor \frac{d}{2} \rfloor + 1$ have depth at most $(\lfloor \frac{d}{2} \rfloor + 1)(r-1)$ which implies $\mathrm{h}_d(n) \leq (\lfloor \frac{d}{2} \rfloor + 1)(r-1)$. Choosing $r := \mathrm{r}_d(n) + 1$ yields

**Corollary 2.3.** $(\lfloor \frac{d}{2} \rfloor + 1) \cdot \mathrm{r}_d(n) \geq \mathrm{h}_d(n)$ *and* $(\lfloor \frac{d}{2} \rfloor + 1) \cdot \mathrm{r}_d^c(n) \geq \mathrm{h}_d^c(n)$.

An identical argument applies to lower half-spaces:

**Corollary 2.4.** $(\lfloor \frac{d}{2} \rfloor + 1) \cdot \mathrm{lr}_d(n) \geq \mathrm{lh}_d(n)$ *and* $(\lfloor \frac{d}{2} \rfloor + 1) \cdot \mathrm{lr}_d^c(n) \geq \mathrm{lh}_d^c(n)$.

## 2.2 Pinning simplices

Similarly, we investigate the relationship between half-space depth and pinning simplices. Recall the definition of the latter:

**Pinning simplices.** What is the largest number $\mathrm{s}_d(n)$ such that every set $P$ of $n$ points in $\mathbb{R}^d$ in general position contains a subset $X \subseteq P, |X| \leq \lfloor \frac{d}{2} \rfloor + 1$ and $\mathrm{s}_d(n)$ other pairwise disjoint subsets $A_1, \ldots, A_{\mathrm{s}_d(n)} \subseteq P \setminus X$ such that the affine hull of $X$ intersects each of the simplices $\operatorname{conv}(A_i)$?

Our earlier observations assert

$$\mathrm{h}_d^c(n) \geq \mathrm{s}_d^c(n).$$

Can we upper bound $\mathrm{h}_d(n)$ in terms of the $\mathrm{s}_d(n)$ as well? As for stabbing centre regions, this is possible and we will prove the following bound:

$$\lceil \tfrac{d}{2} \rceil \cdot \mathrm{s}_d(n) \geq \mathrm{h}_d(n) \quad \text{and} \quad \lceil \tfrac{d}{2} \rceil \cdot \mathrm{s}_d^c(n) \geq \mathrm{h}_d^c(n).$$

We will see the proof in a moment, but first let us give a short preview. Consider a set $X$ of depth $r$. One can orthogonally project the set $X$ to an orthogonal complement of $\operatorname{aff}(X)$ of dimension $d' := d - \dim(\operatorname{aff}(X))$. The depth cannot increase under this projection and by a greedy argument, we obtain $\lceil r/d' \rceil$ vertex-disjoint simplices containing the image of $X$ (which is a single point). Returning to the original space, the corresponding simplices (of dimension $d'$) are intersected by $\mathrm{aff}(X)$. This will conclude the proof. We also present an improvement for $d' = 2$.

**Lemma 2.5.** *Let $P$ be a set of $n$ points in $\mathbb{R}^d$ and $x \in \mathbb{R}^d \setminus P$ a point of depth* $\mathrm{depth}_P^{\mathcal{H}}(x) \geq k$. *Furthermore, assume that $P \cup \{x\}$ is in general position. Then there are $m := \lceil \frac{k}{d} \rceil$ pairwise disjoint subsets $A_1, \ldots, A_m \subseteq P$ each of cardinality $d+1$, such that*

$$x \in \bigcap_{i=1}^{m} \mathrm{int}(\mathrm{conv}(A_i)).$$

*Proof.* We proceed by induction on $k$. For $k = 0$ the statement holds trivially. Otherwise, $x \in \mathrm{int}(\mathrm{conv}(P))$ and by Carathéodory's theorem there is a set $A_m \subseteq P$ of cardinality $|A_m| = d+1$ such that $x \in \mathrm{int}(\mathrm{conv}(A_m))$. Then, by the general position assumption, every half-space with $x$ on its defining hyperplane contains at most $d$ points of $A_m$ and hence, $\mathrm{depth}_{P \setminus A_m}^{\mathcal{H}}(x) \geq \max\{0, k-d\}$. By the induction hypothesis there are $\lceil \frac{k-d}{d} \rceil = m-1$ disjoint subsets $A_1, \ldots A_{m-1}$ of cardinality $d+1$ in $P \setminus A_m$ containing $x$ in the interior of their convex hulls and hence we have the sets $A_1, \ldots, A_m$ as required. □

This lemma seems to be a little wasteful and indeed, in the plane we know how to do better.

**Lemma 2.6.** *Let $P$ be a set of $n$ points in $\mathbb{R}^2$ and $x \in \mathbb{R}^2 \setminus P$ a point of depth* $\mathrm{depth}_P^{\mathcal{H}}(x) = k \leq \lfloor \frac{n}{3} \rfloor$. *Furthermore, assume that $P \cup \{x\}$ is in general position. Then there are $k$ disjoint subsets $A_1, \ldots, A_k \subseteq P$, each of cardinality 3, with $x$ in the intersection of the interiors of their convex hulls.*

*Proof.* Since $\mathrm{depth}_P^{\mathcal{H}}(x) = k$, there is a half-plane $H$ whose defining line passes through $x$ and which contains exactly $k$ points of $P$. Denote these points $t_1, \ldots, t_k$ in their clockwise angular order around $x$ (see Figure 2.3(a) on the next page). Since every half-plane containing $x$ has to contain at least $k$ points of $P$, we have the following property: If we rotate $H$ clockwise around $x$, we must encounter the $i$th new point $r_i$ before losing the point $t_i$ (see Figure 2.3(b) on the facing page). Similarly, if we rotate $H$ anticlockwise, we must encounter the $i$th new point $l_{k+1-i}$ before losing the point $t_{k+1-i}$ on the other side (see Figure 2.3(c) on the next page). Obviously, the sets $T := \{t_1, \ldots t_k\}$, $R := \{r_1, \ldots r_k\}$ and $L := \{l_1, \ldots l_k\}$ are disjoint since $n \geq 3k$.

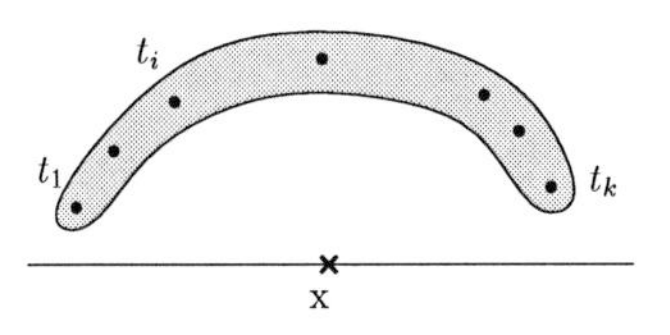


(a) A half-space which witnesses the depth of the point

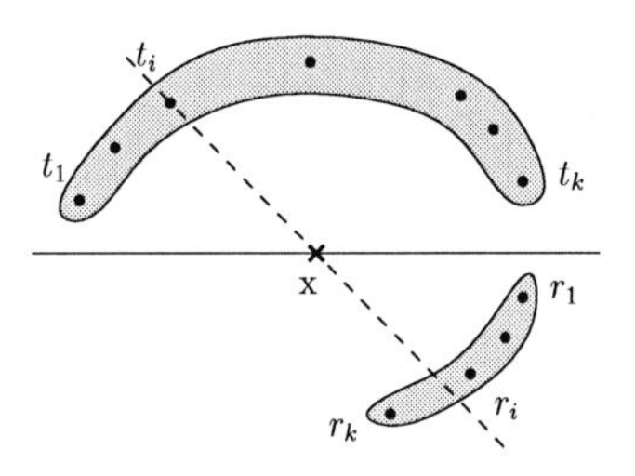


(b) Rotating clockwise encounters $i$ points on the bottom right before losing $i$ points on the top

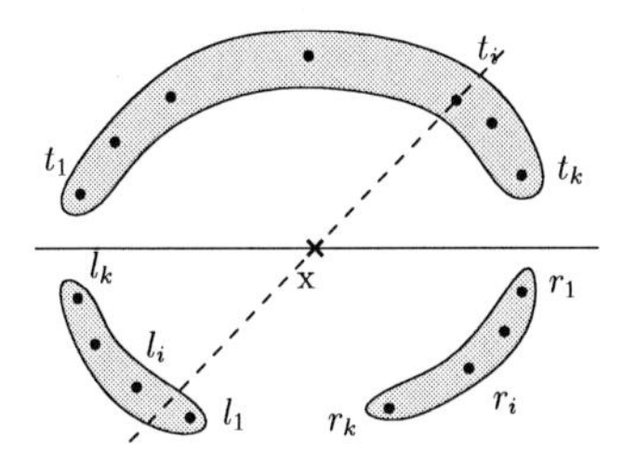


(c) Rotating anticlockwise reveals $i$ points on the bottom left before losing $i$ points on the top

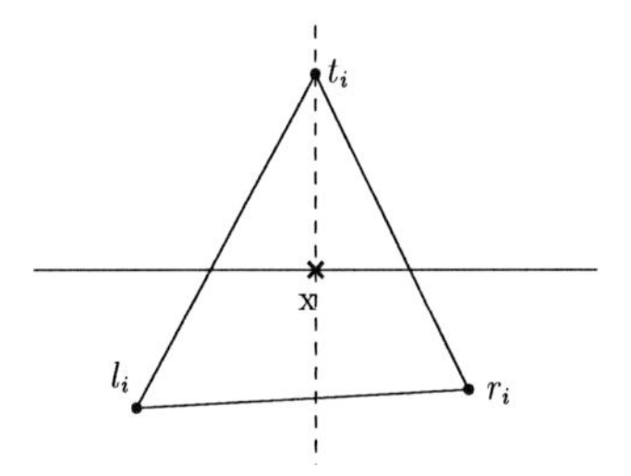


(d) This is sufficient for proving that the triangles $t_i r_i l_i$ contain $x$

Figure 2.3: Proof of Lemma 2.6 in pictures.

The two conditions above imply that (i) the point $r_i$ lies on the left of the oriented line $t_i x$, (ii) the point $l_i$ lies on the right of the oriented line $t_i x$ and (iii) the points $r_i, l_i$ lie on the opposite side of the line defining $H$ from $t_i$. From this, we can deduce that $x$ lies in the triangles $t_i r_i l_i$ (see Figure 2.3(d)) and consequently we get $k$ disjoint sets $A_i := \{t_i, r_i, l_i\}$ containing $x$ in their convex hulls. $\square$

An open question is whether a similar improvement is also possible in higher dimensions.

We can also apply these two lemmata if we have a deep set of cardinality greater than one. We start with a set $X \subseteq P$ of $\text{depth}_P^{\mathcal{H}}(X) \geq k$ and $|X| = d' + 1$. An orthogonal projection $\pi$ onto a $(d - d')$-dimensional orthogonal complement of $\operatorname{aff}(X)$ maps $X$ onto one point $\pi(X)$. It is easy to see that this projection can only increase the depth of $X$ (as pre-images of half-spaces in $\pi(\mathbb{R}^d)$ are half-spaces in $\mathbb{R}^d$) and hence $\operatorname{depth}^{\mathcal{H}}_{\pi(P)}(\pi(X)) \geq k$. Applying one of the lemmata yields $k'$ (depending on the lemma used) disjoint subsets $\pi(A_1), \ldots, \pi(A_{k'})$, each containing $\pi(X)$ in its convex hull. Returning to the pre-images, $\operatorname{aff}(X)$ intersects each of the convex hulls $\operatorname{conv}(A_1), \ldots, \operatorname{conv}(A_{k'})$.

This yields the following corollaries for the pinning simplices problem:

**Corollary 2.7.** $\lceil \frac{d}{2} \rceil \cdot \mathrm{s}_d(n) \geq \mathrm{h}_d(n)$ *and* $\lceil \frac{d}{2} \rceil \cdot \mathrm{s}^{\mathrm{c}}_d(n) \geq \mathrm{h}^{\mathrm{c}}_d(n)$.

**Corollary 2.8.** *For* $d \in \{3, 4\}$ *we have (assuming* $n/3 - 1 \geq \mathrm{h}_d(n)$*)*

$$\mathrm{s}_d(n) \geq \mathrm{h}_d(n) \quad \textit{and} \quad \mathrm{s}^{\mathrm{c}}_d(n) \geq \mathrm{h}^{\mathrm{c}}_d(n).$$

**Remark 2.9.** *Here, it is crucial for the projection step that* $\mathrm{s}_d(n)$ *is defined using the affine rather than the convex hull of* $X$.

## 2.3 Intersecting partitions

The last problem we look into is the intersecting partitions problem and its relationship to the half-space depth problem. Recall the problem description first:

**Intersecting partition.** What is the largest number $\mathrm{p}_d(n)$ such that every set $P$ of $n$ points in $\mathbb{R}^d$ in general position contains $\mathrm{p}_d(n) + 1$ pairwise disjoint subsets $X, A_1 \ldots, A_{\mathrm{p}_d(n)} \subseteq P$ satisfying $|X| \leq \lfloor \frac{d}{2} \rfloor + 1$ and

$$\operatorname{aff}(X) \cap \bigcap_{i=1}^{\mathrm{p}_d(n)} \operatorname{conv}(A_i) \neq \emptyset?$$

From the discussion in the introductory chapter, we already know

$$\mathrm{h}^{\mathrm{c}}_d(n) \geq \mathrm{s}^{\mathrm{c}}_d(n) \geq \mathrm{p}^{\mathrm{c}}_d(n).$$

We again ask the usual question: can we upper bound $\mathrm{h}^{\mathrm{c}}_d$ or $\mathrm{s}^{\mathrm{c}}_d$ in terms of $\mathrm{p}^{\mathrm{c}}_d$? Surprise, surprise ... yes, we can! We will shortly see that

$$(\lceil \tfrac{d}{2} \rceil + 1) \cdot \mathrm{p}_d(n) \geq \mathrm{s}_d(n) \quad \text{and} \quad (\lceil \tfrac{d}{2} \rceil + 1) \cdot \mathrm{p}^{\mathrm{c}}_d(n) \geq \mathrm{s}^{\mathrm{c}}_d(n).$$

**Lemma 2.10.** *Let* $Y$ *be a* $d'$*-dimensional affine space in* $\mathbb{R}^d$ *and consider* $k$ *sets* $A_1, \ldots, A_k \subseteq \mathbb{R}^d$ *of* $d + 1 - d'$ *points each, such that* $Y \cap \operatorname{conv}(A_i) \neq \emptyset$

*for all the sets $A_i$. Then for every $m \in \mathbb{Z}_0^+$ satisfying $k \geq (m-1)(d'+1)+1$ there are $m$ disjoint index sets $I_1, \ldots, I_m$ with*

$$Y \cap \bigcap_{j=1}^{m} \operatorname{conv}\left(\bigcup_{i \in I_j} A_i\right) \neq \emptyset.$$

In another words, the lemma tells us that we can group the sets whose convex hulls intersect $Y$ so that all the groups then have a common intersection on $Y$.

*Proof.* This is a quite straightforward application of Tverberg's theorem. Since $Y$ has affine dimension $d'$ and the sets $A_i$ consist of $d+1-d'$ points, i.e. their affine spans have dimension $d-d'$, the intersections $Y \cap \operatorname{conv}(A_i)$ have dimension zero and therefore, are points. Denote $a_i := Y \cap \operatorname{conv}(A_i)$ for $i \in [k]$ and $T := \{a_i : i \in [k]\}$. All the points in $T$ lie in the affine space $Y$ of dimension $d'$. By the Tverberg's theorem, if $k \geq (m-1)(d'+1)+1$ then there are $m$ disjoint subsets $R_1, \ldots, R_m$ of $T$ such that

$$\bigcap_{i \in [m]} \operatorname{conv}(R_i) \neq \emptyset.$$

Then the index sets $I_j := \{i : a_i \in R_j\}$ satisfy the conclusion of the lemma. □

A short meditation shows that the assumptions of the lemma do exactly match the situation in the pinning simplices problem (we only need to take $Y := \operatorname{aff}(X)$ and $d' = \lceil \frac{d}{2} \rceil$ so that $\lfloor \frac{d}{2} \rfloor + 1 = d + 1 - d'$). As a consequence, if for some $m$ we have $\mathrm{s}_d(n) \geq (m-1)(\lceil \frac{d}{2} \rceil + 1) + 1$ then also $\mathrm{p}_d(n) \geq m$. Let us look at the contrapositive: consider $m := \mathrm{p}_d(n) + 1$ which means $\mathrm{p}_d(n) < m$. This implies $\mathrm{s}_d(n) < (m-1)(\lceil \frac{d}{2} \rceil + 1) + 1$ and substitution yields $\mathrm{s}_d(n) \leq \mathrm{p}_d(n)(\lceil \frac{d}{2} \rceil + 1)$.

**Corollary 2.11.** $(\lceil \frac{d}{2} \rceil + 1) \cdot \mathrm{p}_d(n) \geq \mathrm{s}_d(n)$ *and* $(\lceil \frac{d}{2} \rceil + 1) \cdot \mathrm{p}_d^{\mathrm{c}}(n) \geq \mathrm{s}_d^{\mathrm{c}}(n)$.

**Remark 2.12.** *Since the lemma holds for any convex set $Y$, one can also deduce the same relationship between the convex variants of the problems, that is, when* $\operatorname{aff}(X)$ *is replaced by* $\operatorname{conv}(X)$ *in their respective definitions.*

## 2.4 Summary

There have been too many notions and too many inequalities in this chapter. Let us summarise all the relations between half-space depth and the other problems we have obtained. On the one hand, we have

$$b_{d-1}(n) \geq \mathrm{lh}_d(n) \geq \mathrm{h}_d(n) \quad \text{and} \quad b_{d-1}(n) \geq \mathrm{lh}_d^c(n) \geq \mathrm{h}_d^c(n).$$

For stabbing centre regions, we can say

$$(\lfloor \tfrac{d}{2} \rfloor + 1)\, \mathrm{r}_d(n) \geq \mathrm{h}_d(n) \geq \mathrm{r}_d(n) \quad \text{and} \quad (\lfloor \tfrac{d}{2} \rfloor + 1)\, \mathrm{r}_d^c(n) \geq \mathrm{h}_d^c(n) \geq \mathrm{r}_d^c(n)$$

For pinning simplices and intersecting partitions, we have

$$(\lceil \tfrac{d}{2} \rceil + 1) \cdot \lceil \tfrac{d}{2} \rceil \cdot \mathrm{p}_d(n) \geq \lceil \tfrac{d}{2} \rceil \cdot \mathrm{s}_d(n) \geq \mathrm{h}_d(n) \geq \mathrm{s}_d(n) \geq \mathrm{p}_d(n)$$

and

$$(\lceil \tfrac{d}{2} \rceil + 1) \cdot \lceil \tfrac{d}{2} \rceil \cdot \mathrm{p}_d^c(n) \geq \lceil \tfrac{d}{2} \rceil \cdot \mathrm{s}_d^c(n) \geq \mathrm{h}_d^c(n) \geq \mathrm{s}_d^c(n) \geq \mathrm{p}_d^c(n).$$

We have suggested several variants of the ball depth problem and proved relations between them. So far, none of theses variants has lead to improved bounds but we think that they nonetheless offer interesting viewpoints and possible roads to attack the problem.

**Open problem.** It remains an open question whether some of these relations can be improved. Corollary 2.8 suggests that some improvement should be possible.

# 3 The combinatorics: random sampling

The aim of this chapter is to prove a lower bound for the half-space depth problem in arbitrary dimensions. Together with the statements proved in Chapter 2, this yields lower bounds on ball depth, stabbing centre regions, pinning simplices and intersecting partitions.

In Section 3.1, we consider an extension of the circle containment problem in another direction and prove a dual version of the Neumann-Lara-Urrutia theorem for points and families of pseudo-discs in the plane. The proof demonstrates the main proof method using random sampling but lacks the technical issues which we encounter later.

In Section 3.2, we prove that sets of $n$ points in $\mathbb{R}^d$ in general position have a subset of $\lfloor d/2 \rfloor + 1$ points which has half-space depth on the order of $4n/(6e+1)d^2$ in even dimensions and $4n/(8e+1)d^2$ in odd dimensions. This implies lower bounds for the ball depth problem, stabbing centre regions problem, pinning simplices problem and intersecting partitions problem. The main idea of the proof is to reduce the question to upper bounding the number of objects related to $k$-sets. To upper bound them, one can use random sampling and generalised upper bound theorem.

In Section 3.3, we conclude the chapter by discussing several nuances of the proof, and its improvements in low dimensions.

This is joint work with Shakhar Smorodinsky and Uli Wagner [SSW08].

## 3.1 Pseudo-discs in the plane

In this section, we demonstrate the basic proof technique on the case of discs and pseudo-discs in the plane, where the proof can be carried out without additional technical difficulties.

**Theorem 3.1.** *Let $S$ be a set of $n$ points in $\mathbb{R}^2$ and let $\mathcal{P}$ be a family of pseudo-discs such that the boundary of each member of $\mathcal{P}$ passes through a pair of points in $S$ and each of the $\binom{n}{2}$ possible pairs of points has exactly one element of $\mathcal{P}$ whose boundary passes through it (i.e. there are exactly $\binom{n}{2}$ pseudo-discs). Then there exists a pseudo-disc $D \in \mathcal{P}$ that contains at least $\frac{1}{6e}n - 2$ points of $S$.*

*Proof.* The idea of the proof is the following: We upper bound the number of pseudo-discs with at most $\ell$ points inside. If the upper bound is smaller than the total number of pseudo-discs $\binom{n}{2}$ we can guarantee that there is some pseudo-disc containing more than $\ell$ points. To prove the upper bound, we use the random sampling technique of Clarkson and Shor [CS89] together with an upper bound on the number of empty pseudo-discs by Smorodinsky and Sharir [SS04].

We phrase the problem in the language of the random sampling framework introduced in Section 1.4: The points in $S$ are the objects, the pseudo-discs are the configurations, and incidences, i.e. a point lying in the interior of a pseudo-disc, are the conflicts. Then $\mathcal{P}_{\leq \ell}(S)$ is the set of pseudo-discs containing at most $\ell$ points in their interior.

The upper bound from the random sampling framework (1.2) yields:

$$|\mathcal{P}_{\leq \ell}(S)| \leq \frac{\mathbf{E}\,|\mathcal{P}'_0(S')|}{p^2(1-p)^\ell}, \tag{3.1}$$

where $S'$ is a random sample of $S$ (each point chosen independently with probability $p$) and $\mathcal{P}'$ the set of pseudo-discs induced by $S'$.

Sharir and Smorodinsky [SS04, Lemma 3.4] showed that for any configuration of points and pseudo-discs as above, we have

$$|\mathcal{P}_0(S)| \leq 3n,$$

and consequently, we have in expectation

$$\mathbf{E}[|\mathcal{P}_0(S')|] \leq 3pn. \tag{3.2}$$

Substituting this into the upper bound (3.1) results in:

$$|\mathcal{P}_{\leq\ell}(S)| \leq 3p^{-1}(1-p)^{-\ell}n. \tag{3.3}$$

Setting $p := \frac{1}{\ell+1}$ and using $1 + x \leq e^x$ yields:

$$|\mathcal{P}_{\leq\ell}(S)| \leq 3(\ell+1)\left(\frac{\ell}{\ell+1}\right)^{-\ell} n = 3(\ell+1)\left(1+\frac{1}{\ell}\right)^{\ell} n \leq 3e(\ell+1)n. \tag{3.4}$$

We want to show that there is some pseudo-disc with more than $\ell$ conflicts which is guaranteed if there are fewer than $\binom{n}{2}$ pseudo-discs in $\mathcal{P}_{\leq\ell}(S)$. For that we need:

$$3e(\ell+1)n < \binom{n}{2}. \tag{3.5}$$

This implies $n - 1 > 6e(\ell + 1)$ and choosing $\ell = \frac{n-1}{6e} - 2$ is sufficient. □

Note that Theorem 3.1 implies $b_2(n) \geq \frac{n-1}{6e} - 2$ as a special case. Indeed, let $S$ be a set of points in $\mathbb{R}^2$. Then for every pair $A \in \binom{S}{2}$, define $D_A$ to be a disc containing $A$ with the smallest possible intersection* $D_A \cap S$. The family $\mathcal{P} := \{D_A \mid A \in \binom{S}{2}\}$ is a family of pseudo-discs, hence there is a disc $D \in \mathcal{P}$ (which is $D = D_A$ for some pair $A$) containing at least $\frac{1}{6e}n - 2$ points of $S$. By the choice of $D_A$, we know that every disc containing $A$ has to contain at least $\frac{1}{6e}n - 2$ points of $S$.

## 3.2 Lower bounds in arbitrary dimension

In this section, we switch to a general setting of half-space depth and prove a result of a similar nature as Theorem 3.1 in Section 3.1.

The results will be different in odd and even dimensions, hence let us introduce the following function $\mathbf{2}_d$ defined by

$$\mathbf{2}_d = \begin{cases} 1.5 & d \text{ even}, \\ 2 & d \text{ odd}. \end{cases}$$

**Theorem 3.2** (Half-space depth).

$$\mathrm{h}_d(n) \geq \frac{n}{(4e \cdot \mathbf{2}_d + 1)(\lfloor \frac{d}{2} \rfloor + 3)(\lfloor \frac{d}{2} \rfloor + 1)} + O(1).$$

*To fulfil the additional requirement about the boundaries, we simply shrink the disc until both the points are on its boundary.

If we consider lower half-spaces instead of general half-spaces, the expression $4e \cdot \mathbf{2}_d + 1$ in the constant can be improved to $2e \cdot \mathbf{2}_d + 1$.

**Proposition 3.3** (Lower half-space depth).

$$\mathrm{lh}_d(n) \geq \frac{n}{(2e \cdot \mathbf{2}_d + 1)(\lfloor \frac{d}{2} \rfloor + 3)(\lfloor \frac{d}{2} \rfloor + 1)} + O(1).$$

In the special case $d = 3$, a key step in our proof (tight lower bounds on the number of $(\leq k)$-sets) can be refined to give an improved constant of $1/(8e)$ (respectively, of $1/6e$, for points in convex position). The factors improve, if larger sets $X$ are considered instead (with slightly refined analysis). We discuss these issues in more detail in the next section. We also note the following immediate consequence of the theorem, combined with Corollary 2.3 on page 31:

**Corollary 3.4** (Stabbing centre regions).

$$\mathrm{r}_d(n) \geq \frac{n}{(4e \cdot \mathbf{2}_d + 1)(\lfloor \frac{d}{2} \rfloor + 3)(\lfloor \frac{d}{2} \rfloor + 1)^2} + O(1)$$

Similarly, we obtain the following two corollaries immediately from Corollary 2.7 on page 34 and Corollary 2.11 on page 35.

**Corollary 3.5** (Pinning simplices).

$$\mathrm{s}_d(n) \geq \frac{n}{(4e \cdot \mathbf{2}_d + 1)(\lfloor \frac{d}{2} \rfloor + 3)(\lfloor \frac{d}{2} \rfloor + 1)\lceil \frac{d}{2} \rceil} + O(1).$$

**Corollary 3.6** (Intersecting partitions).

$$\mathrm{p}_d(n) \geq \frac{n}{(4e \cdot \mathbf{2}_d + 1)(\lfloor \frac{d}{2} \rfloor + 3)(\lfloor \frac{d}{2} \rfloor + 1)\lceil \frac{d}{2} \rceil(\lceil \frac{d}{2} \rceil + 1)} + O(1).$$

Last but not least, by the paraboloidal lifting map discussed in the beginning of Section 1.2 and Proposition 3.3 we also get:

**Corollary 3.7** (Ball depth).

$$\mathrm{b}_d(n) \geq \frac{n}{(2e \cdot \mathbf{2}_{d+1} + 1)(\lfloor \frac{d+1}{2} \rfloor + 3)(\lfloor \frac{d+1}{2} \rfloor + 1)} + O(1).$$

The proofs of Theorem 3.2 and Proposition 3.3 are almost identical and we prove them by one "meta-proof" where everything is indexed by a family of half-spaces $\mathcal{R} \in \{\mathcal{H}, \mathcal{H}^-\}$ — please do not let that frighten you or even discourage you from further reading.

Let us sketch the structure of the proof of Theorem 3.2 before proceeding. As in the case of pseudo-discs, we try to upper bound the number of sets $X \subseteq P$ of cardinality $\lfloor \frac{d}{2} \rfloor + 1$ and depth at most $\lfloor \frac{d}{2} \rfloor + 1 + \ell$. As long as there are provably fewer than $\binom{n}{\lfloor \frac{d}{2} \rfloor + 1}$ of these sets, we won the jackpot since some subsets $X$ must have depth higher than $\ell$. For each set $X$ of $\lfloor \frac{d}{2} \rfloor + 1$ points, there is some half-space $H$ witnessing its depth, i.e. $|H \cap P| = \mathrm{depth}_P^{\mathcal{H}}(X)$ and $X \subseteq H$. In this setting, one can use random sampling which reduces the problem to upper bounding the number of $k$-sets of $P$ for which we can use the generalised upper bound theorem [Wag06].

### 3.2.1 $k$-set estimates

Before we prove the theorem, we will need technical estimates of two quantities related to $k$-sets in order to be able to use the bounds obtained by the random sampling.

For any finite point set $P \subseteq \mathbb{R}^d$ and a set of half-spaces $\mathcal{R} \in \{\mathcal{H}, \mathcal{H}^-\}$, let $a_{k,\leq\ell}^{\mathcal{R}}(P)$ (and $a_{k,\ell}^{\mathcal{R}}(P)$) be the number of $k$-element subsets $A \subseteq P$ for which there is a half-space in $\mathcal{R}$ containing $A$ and at most $\ell$ (exactly $\ell$) points of $P \setminus A$. Note that for any point set $X$,

$$a_k^{\mathcal{R}}(P) := a_{k,0}^{\mathcal{R}}(P)$$

is the number of *$k$-sets* or the number of *lower $k$-sets* of $P$ (for $\mathcal{R} = \mathcal{H}$ and $\mathcal{R} = \mathcal{H}^-$, respectively), i.e. of $k$-element subsets $Y \subseteq P$ such that $Y = P \cap H$ for some half-space $H$ (respectively, lower half-space $H$). We remark that $a_{k,0}^{\mathcal{R}}(P)$ is related, but in general not equal, to the number of $(k-1)$-dimensional faces (respectively, lower faces) of the convex hull $\mathrm{conv}(P)$. This number is notoriously difficult to estimate in general. In a first step, we use the obvious estimate $a_k^{\mathcal{R}}(P) \leq a_{\leq k}^{\mathcal{R}}(P) := a_1^{\mathcal{R}}(P) + \ldots + a_k^{\mathcal{R}}(P)$. In our case, we will be interested in $k = \lfloor d/2 \rfloor + 1$ which only depends on $d$, not on $n$, so we do not lose too much in this step. The second step is to use worst-case estimates for $a_{\leq k}^{\mathcal{R}}(P)$.

**Remark 3.8.** *This might be a place, where we lose up to a factor of $\lfloor \frac{d}{2} \rfloor + 1$ in our final bounds.*

This was one of the first problems to which Clarkson and Shor applied their random sampling technique. Their bounds are tight up to a constant depending only on $d$, but for very small $k$, as in our case, this constant is quite large (exponential in $d$). Sharper bounds were proved by Wagner [Wag06, Corollary 2.5]:

**Lemma 3.9** ([Wag06]). *Let $P$ be a set of $n$ points in $\mathbb{R}^d$ and let $C_{n,d}$ denote a set of $n$ points on the moment curve in $\mathbb{R}^d$ (the choice of the points does not matter). Let $\mathcal{R} \in \{\mathcal{H}, \mathcal{H}^-\}$. Then[1] for any integer $1 \leq k \leq n-1$,*

$$a^{\mathcal{R}}_{\leq k}(P) \leq c_{\mathcal{R}} \cdot \sum_{i=1}^{d} a^{\mathcal{H}}_{\leq k}(C_{n,i}), \tag{3.6}$$

*where $c_{\mathcal{R}} = 2$ if $\mathcal{R} = \mathcal{H}^-$ and $c_{\mathcal{R}} = 4$ if $\mathcal{R} = \mathcal{H}$.*

For points on the moment curve, the numbers $a^{\mathcal{H}}_k$ were computed exactly by Andrzejak and Welzl [AW03, Corollary 5.2]:

**Lemma 3.10** ([AW03]). *Let $C_{n,d}$ be a set of $n$ points on the moment curve in $\mathbb{R}^d$. Then, for $1 \leq k \leq n-1$, the number of $k$-sets in $C_{n,d}$ equals*

$$a^{\mathcal{H}}_k(C_{n,d}) = \sum_{s=0}^{d} \left[ \binom{k-1}{\lfloor \frac{s-1}{2} \rfloor} \binom{n-k-1}{\lfloor \frac{s}{2} \rfloor} + \binom{k-1}{\lfloor \frac{s}{2} \rfloor} \binom{n-k-1}{\lfloor \frac{s-1}{2} \rfloor} \right] \tag{3.7}$$

The left term contains $\binom{n-k-1}{t}$ for $s \in \{2t, 2t+1\}$ and the right term contains $\binom{n-k-1}{t}$ for $s \in \{2t+1, 2t+2\}$. Regrouping the terms of (3.7) according to the binomial coefficient $\binom{n-k-1}{t}$ and using Pascal's rule three times yields the following upper bound:

$$\begin{aligned} a^{\mathcal{H}}_k(C_{n,d}) &\leq \sum_{t=0}^{\lfloor \frac{d}{2} \rfloor} \binom{n-k-1}{t} \cdot \left( \binom{k-1}{t-1} + 2\binom{k-1}{t} + \binom{k-1}{t+1} \right) \\ &= \sum_{t=0}^{\lfloor \frac{d}{2} \rfloor} \binom{k+1}{t+1} \cdot \binom{n-k-1}{t} \leq \sum_{t=0}^{\lfloor \frac{d}{2} \rfloor} \binom{k+1}{t+1} \cdot \binom{n}{t}. \end{aligned}$$

[1] The results in [Wag06] are phrased in terms of levels in arrangements of affine hyperplanes or half-spaces. By standard point-hyperplane duality, the $k$-sets of a set of $n$ points in $\mathbb{R}^d$ correspond to the the $d$-dimensional cells at lower level $k$ plus the $d$-dimensional cells at upper level $k$ in an arrangement of $n$ non-vertical affine hyperplanes in $\mathbb{R}^d$. The fact that we have to consider lower and upper levels separately leads to the factor of 4 for arbitrary half-spaces in Lemma 3.9, instead of the factor of 2 in the quoted corollary.

To estimate the $a_{\leq k}(C_{n,d})$ we sum up all $a_i(C_{n,d})$ for $i \leq k$ and use Fact 1.8:

$$\begin{aligned} a^{\mathcal{H}}_{\leq k}(C_{n,d}) &= \sum_{i=0}^{k} a_i(C_{n,d}) \leq \sum_{i=0}^{k} \sum_{t=0}^{\lfloor \frac{d}{2} \rfloor} \binom{i+1}{t+1} \cdot \binom{n}{t} \\ &= \sum_{t=0}^{\lfloor \frac{d}{2} \rfloor} \binom{n}{t} \sum_{i=0}^{k} \binom{i+1}{t+1} = \sum_{t=0}^{\lfloor \frac{d}{2} \rfloor} \binom{n}{t} \binom{k+2}{t+2}. \end{aligned}$$

Combining with Lemma 3.9, we obtain an upper bound for arbitrary point set:

$$a^{\mathcal{R}}_{\leq k}(P) \leq c_{\mathcal{R}} \cdot \sum_{i=1}^{d} a_{\leq k}(C_{n,i}) \leq c_{\mathcal{R}} \cdot \sum_{i=1}^{d} \sum_{t=0}^{\lfloor \frac{i}{2} \rfloor} \binom{n}{t} \binom{k+2}{t+2}.$$

Carefully switching the order of summation yields

$$a^{\mathcal{R}}_{\leq k}(P) \leq c_{\mathcal{R}} \cdot \sum_{t=0}^{\lfloor \frac{d}{2} \rfloor} \sum_{i=2t}^{d} \binom{n}{t} \binom{k+2}{t+2} \leq c_{\mathcal{R}} \cdot \sum_{t=0}^{\lfloor \frac{d}{2} \rfloor} \binom{n}{t} \binom{k+2}{k-t} (d-2t+1)$$

The binomial coefficients turn out to be inconvenient for further calculations and hence, we upper bound them by a geometric series (recall that $t \leq \lfloor \frac{d}{2} \rfloor$ and furthermore, we only need $k = \lfloor \frac{d}{2} \rfloor + 1$; further explanation is given below):

$$\binom{k+2}{k-t}(d-2t+1) \leq (k+2)^{k-t} \cdot \frac{d-2t+1}{(k-t)!} \leq (k+2)^{k-t} \cdot \mathbf{2}_d$$

To see the last inequality, we need to find the maximum of the fraction $\frac{d-2t+1}{(k-t)!}$. The first few values of the fraction for $t = \lfloor \frac{d}{2} \rfloor, \lfloor \frac{d}{2} \rfloor - 1, \ldots$ are $\frac{2}{1}, \frac{4}{2}, \frac{6}{6}, \frac{8}{24}$ for $d$ odd and $\frac{1}{1}, \frac{3}{2}, \frac{5}{6}, \frac{7}{24}$ for $d$ even. It is easy to see that the maximum of the fraction is attained among these first few values and it is 1.5 in even dimensions and 2 in odd dimensions. Using this bound, we can conclude:

**Lemma 3.11.** *Let $P$ be a set of $n$ points in $\mathbb{R}^d$. Then*

$$a^{\mathcal{R}}_{k}(P) \leq c_{\mathcal{R}} \cdot \sum_{t=0}^{\lfloor \frac{d}{2} \rfloor} \binom{n}{t} (k+2)^{k-t} \mathbf{2}_d \tag{3.10}$$

This bound turns out to be good enough for small values of $k - \lfloor \frac{d}{2} \rfloor$ (which is our case). However, if we want to do the calculations for larger $k$ (that is, when we are looking on larger sets $X$ of high depth), where $k - \lfloor \frac{d}{2} \rfloor$ is proportional to $\lfloor \frac{d}{2} \rfloor$, the trivial upper bound

$$\binom{k+2}{k-t} \leq 2^{k+2} \tag{3.11}$$

fares much better.

### 3.2.2 Random sampling

Although the main structure of our proof of Theorem 3.2 (and Proposition 3.3) remains the same as in Theorem 3.1 (Clarkson-Shor argument combined with an appropriate upper bound on the number of certain configurations), most of the work here comes from technical details (we have already seen part of them in the previous subsection).

*Proof of Theorem 3.2 and Proposition 3.3.* We will upper bound the number of subsets $X \subseteq P, |X| = k := \lfloor \frac{d}{2} \rfloor + 1$ of depth at most $k + \ell$, i.e. the values $a^{\mathcal{R}}_{k,\leq \ell}(P)$. As long as we can show that the upper bound is smaller than the number of all such subsets $\binom{n}{k}$, this will guarantee that there is some subset $X$ of depth at least $\ell$.

To express our problem in the random sampling framework, we need objects, configurations and conflicts. The set of objects will be the set of points $P$. For each subset $X \subseteq P$ of cardinality $|X| = k$ there is some half-space $H_X \in \mathcal{R}$ witnessing the depth of $X$, i.e. with $X \subseteq H_X$ and $|H_X \cap P| = \text{depth}^{\mathcal{R}}_P(X)$. These half-spaces are the configurations. Each of them is defined by $k$ objects. Even if, for some reason, geometrically identical half-space is picked for two different subsets $X$, we treat them as different (one can think of the set $X$ as being part of the description of the configuration). A point $p \in P$ is in conflict with the half-space $H_X$ if $p \notin X$ and $p \in H_X$. In this setting, $H_X \cap P$ is the depth of $X$. However, we will perform random sampling and in the sample, the half-spaces do not necessarily have to reflect the depth any more. Nevertheless, they give an upper bound which is all we need (recall that we will only look at configurations of weight 0 and definitely, all the configurations whose half-space has no conflicts fulfils this condition).

Let us keep in mind that we want to compare the upper bound with $\binom{n}{k}$ in the end: we will do the calculations in such a way that we are prepared for that.

The random sampling framework (1.2) yields

$$a^{\mathcal{R}}_{k,\leq \ell}(P) \leq \mathbf{E}[a^{\mathcal{R}}_{k,0}(P')]p^{-k}(1-p)^{-\ell}, \tag{3.12}$$

where $P'$ is a random subset of $P$ obtained by choosing each point independently with probability $p$, for a parameter $0 < p < 1$ that will be specified later.

Note that the cardinality of the random subset $P'$ of $P$ has the binomial distribution $\mathrm{B}(n, p)$. We apply Lemma 3.11 and Fact 1.9 to the random subset $P'$ and obtain

$$\mathbf{E}[a_k^{\mathcal{R}}(P')] \leq c_{\mathcal{R}} \cdot \sum_{t=0}^{\lfloor \frac{d}{2} \rfloor} p^t \binom{n}{t} (k+2)^{k-t} \mathbf{2}_d. \tag{3.13}$$

Substituting this into (3.12) yields

$$a_{k,\leq \ell}^{\mathcal{R}}(P) \leq c_{\mathcal{R}} \cdot p^{-k} (1-p)^{-\ell} \cdot \sum_{t=0}^{\lfloor \frac{d}{2} \rfloor} p^t \binom{n}{t} (k+2)^{k-t} \mathbf{2}_d.$$

We conveniently set $p := \ell^{-1}$ (this is not too far from the best value of $p$) and use the fact that $1 + x \leq e^x$ for every $x \in \mathbb{R}$:

$$\begin{aligned}
a_{k,\leq \ell}^{\mathcal{R}}(P) &\leq \mathbf{2}_d c_{\mathcal{R}} \ell^k (1 - \ell^{-1})^{-\ell} \cdot \sum_{t=0}^{\lfloor \frac{d}{2} \rfloor} p^t \binom{n}{t} (k+2)^{k-t} \\
&\leq \mathbf{2}_d c_{\mathcal{R}} e \ell^k \cdot \sum_{t=0}^{\lfloor \frac{d}{2} \rfloor} \binom{n}{t} \frac{(k+2)^{k-t}}{\ell^t} \\
&\leq \mathbf{2}_d c_{\mathcal{R}} e \ell^k (k+2)^k \cdot \sum_{t=0}^{\lfloor \frac{d}{2} \rfloor} \left( \frac{n}{(k+2)\ell} \right)^t \frac{1}{t!}.
\end{aligned}$$

Multiplying by $k!/k!$ and upper bounding $k!/t! \leq k^{k-t}$ results in

$$a_{k,\leq \ell}^{\mathcal{R}}(P) \leq (k!)^{-1} \mathbf{2}_d c_{\mathcal{R}} e \ell^k (k+2)^k \cdot \sum_{t=0}^{\lfloor \frac{d}{2} \rfloor} \left( \frac{n}{(k+2)\ell} \right)^t k^{k-t},$$

which leaves us with the inequality

$$a_{k,\leq \ell}^{\mathcal{R}}(P) \leq (k!)^{-1} \mathbf{2}_d c_{\mathcal{R}} e ((k+2)k\ell)^k \cdot \sum_{t=0}^{\lfloor \frac{d}{2} \rfloor} \left( \frac{n}{k(k+2)\ell} \right)^t.$$

If $\frac{k(k+2)\ell}{n} \leq \mu_k < 1$ for some constant $\mu_k > 0$, we denote $\nu_k := \frac{1}{1-\mu_k}$ and we upper bound the sum of the geometric series starting at $t = \lfloor \frac{d}{2} \rfloor$ (with $t$ decreasing) by

$$a^{\mathcal{R}}_{k,\leq\ell}(P) \leq (k!)^{-1}\mathbf{2}_d c_{\mathcal{R}} e(k(k+2)\ell)^k \cdot \left(\frac{n}{k(k+2)\ell}\right)^{\lfloor \frac{d}{2} \rfloor} \nu_k$$

which simplifies to

$$a^{\mathcal{R}}_{k,\leq\ell}(P) \leq (k!)^{-1} n^{\lfloor \frac{d}{2} \rfloor} \mathbf{2}_d c_{\mathcal{R}} e(k(k+2)\ell)^{k-\lfloor \frac{d}{2} \rfloor} \nu_k.$$

Now is the right time to say that we are looking for $\ell$ such that this upper bound is smaller than $\binom{n}{k}$. The highest such value is obtained by finding $\ell$ which makes the upper bound equal to $\binom{n}{k}$ and subtracting one afterwards. For algorithmic implications (we want to see how does the target $\ell$ change if we want a certain fraction of the subsets to have depth higher than $\ell$ for sampling purposes), we will introduce an artificial constant $\alpha$ and require that the upper bound is equal $\alpha\binom{n}{k}$. You can think of it as being one for the sake of this proof.

$$\alpha\binom{n}{k} = (k!)^{-1} n^{\lfloor \frac{d}{2} \rfloor} \mathbf{2}_d c_{\mathcal{R}} e(k(k+2)\ell)^{k-\lfloor \frac{d}{2} \rfloor} \nu_k. \tag{3.15}$$

By pulling the terms to the left we obtain

$$\frac{\alpha \cdot (1+O(n^{-1}))}{\mathbf{2}_d c_{\mathcal{R}} e \nu_k} \left(\frac{n}{k(k+2)}\right)^{k-\lfloor \frac{d}{2} \rfloor} = \ell^{k-\lfloor \frac{d}{2} \rfloor}$$

Recall that $k - \lfloor \frac{d}{2} \rfloor = 1$ which implies:

$$\frac{\alpha}{\mathbf{2}_d c_{\mathcal{R}} e \nu_k} \cdot \frac{n}{k(k+2)} + O(1) = \ell$$

The last ingredient of the bound, which is still missing, is to determine some of $\mu_k, \nu_k$ such that $\frac{k(k+2)\ell}{n} \leq \mu_k < 1$ . It is sufficient to have

$$\frac{\alpha}{\mathbf{2}_d c_{\mathcal{R}} e \nu_k} \cdot \frac{n}{k(k+2)} \cdot \frac{k(k+2)}{n} + O(n^{-1}) \leq \mu_k,$$

by simply comparing the lower bound on $\ell$ with the constraint on $\mu_k$. It simplifies to

$$\frac{\alpha}{\mathbf{2}_d c_{\mathcal{R}} e \nu_k} \leq \mu_k + O(n^{-1}).$$

Substituting for $\mu_k$ using $\nu_k = \frac{1}{1-\mu_k} \Rightarrow \mu_k = \frac{\nu_k - 1}{\nu_k}$ results in:

$$\frac{\alpha}{\mathbf{2}_d c_{\mathcal{R}} e} \leq (1 + O(n^{-1}))\nu_k - 1$$

Hence, $\nu_k := (1 + O(n^{-1}))\frac{\mathbf{2}_d c_{\mathcal{R}} e + \alpha}{\mathbf{2}_d c_{\mathcal{R}} e}$ is a feasible choice. Consequently, we have $\frac{\alpha}{\mathbf{2}_d c_{\mathcal{R}} e \nu_k} = \frac{\alpha + O(n^{-1})}{\mathbf{2}_d c_{\mathcal{R}} e + \alpha}$. Now, we can finally substitute to obtain the final bound on $\ell$:

$$\frac{\alpha}{\mathbf{2}_d c_{\mathcal{R}} e + \alpha} \cdot \frac{n}{k(k+2)} + O(1) \leq \ell.$$

□

Note that the estimates in this proof were not always optimal. Our goal was showing the quadratic dependence without introducing more technical issues than necessary.

## 3.3 Thoughts and discussion

The proof presented in the previous section can be improved under certain circumstances which we will discuss now.

**Remark 3.12.** (i) *In $d = 3$, exact upper bounds for the number $e_k$ of so-called ($\leq k$)-*facets[1] *were proved by Welzl [Wel01, Corollary 8]: for every set $P$ of $n$ points in $\mathbb{R}^3$, $e_{\leq k}(P) \leq 2\left[\binom{k+2}{2}n - 2\binom{k+3}{3}\right]$ (and this is attained for point sets in convex position). Together with the linear equation $a_k = (e_{k-1} + e_{k-2})/2 + 2$ valid for any point set in $\mathbb{R}^3$ (see [AW03, Remark 5]), this easily implies $a_2(P) \leq a_{\leq 2}(P) \leq 4n$ for the special case $k = k_3 = 2$. This drastically simplifies the computations for the Clarkson-Shor estimate and yields $\binom{n}{2} = a_{2,\leq \ell}(P) \leq 4np(1-p)^{-\ell}$. Setting $p = 1/(1+\ell)$ as before then shows that there is a pair $A \in \binom{P}{2}$ of* $\mathrm{depth}_P^{\mathcal{H}}(A) \geq \frac{n-1}{8e} - 2$.

(ii) *If, in addition to $d = 3$, we further assume that the point set $P$ is in convex position (which is the case, for instance, for points on the paraboloid $U$ obtained by the lifting map), then a 2-set of $P$ is actually an edge of the convex hull. Thus, $a_k(P) \leq 3n-6$ which yields a further improvement in the constant from $1/8e$ to $1/6e$.*

[1] A $k$-facet of a point set $P$ in general position is a $(d-1)$-dimensional simplex $\sigma$ such that one of the two open half-spaces determined by $\sigma$ contains exactly $k$ points of $P$.

(iii) *As we already hinted in the proof of the Theorem 3.2, one might consider* $k > \lfloor \frac{d}{2} \rfloor + 1$ *in Theorem 3.2. Our method gives improved values of the constant* $\alpha$ *for larger* $k$ *but these are still on the order of* $1/d^2$*. If* $k - \lfloor \frac{d}{2} \rfloor$ *is relatively large (i.e. on the order of* $c \cdot d$ *if* $d$ *is growing as well, but much slower than* $n$*), one can replace the upper bound from Lemma 3.11 by the trivial (3.11). This removes the factor* $k + 2$ *and replaces it by a constant depending on the value* $c$ *from above. Hence, the dependence for larger* $k$ *is only on the order* $1/d$*.*

(iv) *The proof in [RV09] actually shows that, in a set* $P$ *of* $n$ *points in* $\mathbb{R}^3$*, there is a pair* $\{a, b\}$ *of lower half-space depth around* $n/4.73$*. By Corollary 2.8 on page 34 and Corollary 2.11 on page 35 we get a partition of* $P$ *into* $n/9.46$ *parts, one of which is* $\{a, b\}$ *and namely, has cardinality two.*

# 4 Random thoughts

As the name of the chapter might suggest, this is a collection of various thoughts related to the problems and the proofs presented in the previous chapters. There will be two directions we look into: upper bounds on the problems of half-space depth, ball depth and other; the other direction are algorithmic consequences of our lower bound proof.

## 4.1 Upper bounds

We proved lower bounds on the problem of half-space depth and consequently, also obtained lower bounds for ball depth problem, intersecting partitions problem etc. It is natural to ask, how far from the truth these lower bounds can be. Here we give upper bounds to have some comparison. Essentially, all the upper bounds we present are of the order $cn/d$ which indicates, that there is a space for improvement, either for the upper bounds or for the lower bounds.

#### Intersecting partitions

It is well known and easy to see by a dimensionality argument that no set of $n = (r-1)(d+1)$ points in $\mathbb{R}^d$ in general position has a Tverberg partition, i.e.

partition into $r$ disjoint parts whose convex hulls intersect. Indeed, assume to the contrary that $A_1, \ldots A_r$ are sets in some valid partition. Then $\sum_{i=1}^{r} |A_i| = n$ which implies

$$\begin{aligned} \operatorname{codim}\left(\bigcap_{i=1}^{r} \operatorname{aff}(A_i)\right) &= \sum_{i=1}^{r} \operatorname{codim}(\operatorname{aff}(A_i)) \\ &= \sum_{i=1}^{r} ((d+1) - |A_i|) \\ &= r(d+1) - n = d+1. \end{aligned}$$

If the intersection of the affine hulls has co-dimension $d+1$ then it must have dimension $d - (d+1) = -1$ and consequently, is empty. Therefore, the intersection of the convex hulls has to be empty as well (since $\operatorname{conv}(X) \subseteq \operatorname{aff}(X)$).

Hence, a partition of a set $P$ of $n$ points in general position is only possible into $\frac{n-1}{d+1} + 1$ or fewer parts with intersecting convex hulls. We can conclude:

$$\mathrm{p}_d(n) \leq \frac{n-1}{d+1} + 1 \quad \text{and} \quad \mathrm{p}_d^c(n) \leq \frac{n-1}{d+1} + 1.$$

### Stabbing centre regions

We have already seen, that the $r$-centre regions and lower $r$-centre region can be empty for $r > \frac{n}{d+1}$ and $r > \frac{n}{d}$, respectively (additionally, both can be attained in convex position). This implies

$$\mathrm{r}_d(n) \leq \frac{n}{d+1} + 1 \quad \text{and} \quad \mathrm{r}_d^c(n) \leq \frac{n}{d+1} + 1$$

and

$$\mathrm{lr}_d(n) \leq \frac{n}{d} + 1 \quad \text{and} \quad \mathrm{lr}_d^c(n) \leq \frac{n}{d} + 1.$$

For stabbing lower centre regions, there is a simple construction which improves this trivial upper bound to $\lceil \frac{n}{d+1} \rceil$. In fact, it even shows that there is an $n$ point set on *a paraboloid* such that for $r = \lceil \frac{n}{d+1} \rceil + 1$, $\operatorname{conv} A \cap C_r^{\mathcal{H}^-}(P) = \emptyset$ for all $(d-1)$-element subsets.

For simplicity, suppose that $n = t(d+1)$. Consider the vertex set of a regular $(d-1)$-simplex centred at the origin $o$ in $\mathbb{R}^{d-1}$ and replace each vertex $v_i$ by a sufficiently small cluster $V_i$ of $t$ points. Similarly, replace $o$ by a small cluster of $t$ points. Let $S_i = \hat{V}_i$ be the lifting of $V_i$ to the paraboloid and let $P := S_0 \cup S_1 \cup \ldots \cup S_d$ (the lifted clusters $\hat{V}_i$ are located at the vertices of

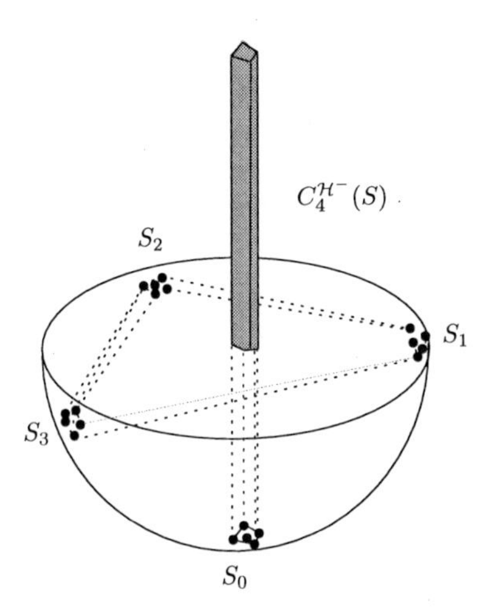


Figure 4.1: A lower centre region not intersected by a segment.

a $d$-simplex that is "pointing downward"). Consider $C := C_r^{\mathcal{H}^-}(P)$, with $r = \frac{n}{d+1}+1$. Clearly, $C$ is contained in the vertical cylinder $\operatorname{conv}(V_0)\times\mathbb{R}$. Moreover, $C$ lies above the lower convex hull of the "upper" clusters $S_1 \cup \ldots \cup S_d$. It is easy to see that $\operatorname{conv} A \cap C = \emptyset$ for all $A \in \binom{P}{d-1}$. Consequently, we have a slight improvement

$$\operatorname{lr}_d(n) \leq \left\lceil \frac{n}{d+1} \right\rceil + 1 \quad \text{and} \quad \operatorname{lr}_d^c(n) \leq \left\lceil \frac{n}{d+1} \right\rceil + 1.$$

### Half-space depth

Recall the moment curve $\gamma(t) = \{t, t^2, \ldots, t^d\}, t \geq 0$, in $\mathbb{R}^d$. Every hyperplane intersects the moment curve in at most $d$ points (since the values of parameter $t$ corresponding to the intersections are roots of a polynomial of degree at most $d$) and for any $d$ points on the moment curve, there is a hyperplane passing through them (as generally, for any $d$ points there is a hyperplane passing through those points). Furthermore, any set of $n$ points on the moment curve is in convex position and $\gamma(t)$ lies above any fixed hyperplane as $t \to \infty$.

Let $0 < t_1 < \ldots < t_n$ and denote $C := \{\gamma(t_i) \colon i \in [n]\}$. We say, that a point $\gamma(s)$ is to the left of $\gamma(t)$ if $s < t$. Consider any $k := \lfloor \frac{d}{2} \rfloor + 1$ points $\gamma(s_1), \ldots, \gamma(s_k) \in C$ where $s_1 < \ldots < s_k$ and set $n' := n - k$. We try to find a half-space with the points $\gamma(s_1), \ldots, \gamma(s_k)$ on its boundary which does not

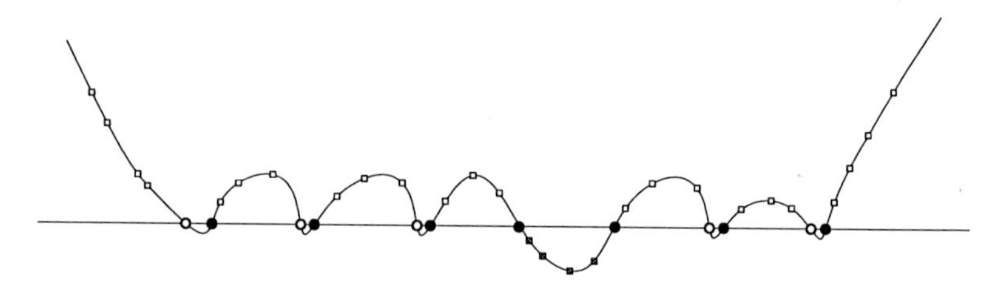

Figure 4.2: Choosing a lower half-space through the moment curve: white circles are the free choice.

contain too many of the remaining $n'$ points of $C$. The points $\gamma(s_i)$ split the moment curve into $k+1$ parts and our strategy will be finding a half-space containing points from only one of these parts. By the pigeonhole principle, if we have $\ell$ half-spaces containing disjoint subsets of $n'$ points then one of them has to contain at most $n'/\ell$ points.

We have $k$ intersection points fixed, but remaining $d-k = \lceil \frac{d}{2} \rceil - 1$ can be chosen freely. This gives us $k-2$ free intersection points in even dimensions and $k-1$ in odd dimensions. We need to consider four different settings: odd and even dimensions, and lower half-spaces or all half-spaces. Only lower half-spaces in even dimensions gives us the least freedom and we will look at it first. Every other setting will bring us additional "choices" and consequently, a better upper bound.

We show that for every $i \in [2,k]$ there is a lower half-space which only contains those points of $C$ which lying between $\gamma(s_{i-1})$ and $\gamma(s_i)$. We specify the arguments (i.e. $t$ of the point $\gamma(t)$) of intersections of the moment curve with the boundary hyperplane of the half-space (from right to left) as (see Figure 4.2):

$$s_k, s_k - \delta, s_{k-1}, \ldots s_{i+1} - \delta, \mathbf{s_i}, \mathbf{s_{i-1}}, s_{i-2}, s_{i-2} - \delta, \ldots, s_1, s_1 - \delta.$$

Here, $\delta$ is some small positive constant such that $\delta < \min_{\gamma(a),\gamma(b)\in C} |a-b|$. Then the only parts of the moment curve contained in the lower half-space are the ones with parameter values in $(s_j - \delta, s_j), j \neq i$ and $(s_{i-1}, s_i)$. The choice of $\delta$ guarantees that the former ones are empty of points of $C$. We also need to check that we did not specify too many intersection points: we used $k$ points $\gamma(s_i), i \in [k]$ and $k-2$ additional points $\gamma(s_i - \delta), i \in [k] \setminus \{i-1, i\}$. This shows the existence of such half-spaces. It also implies that we have $k-1$ choices of lower half-spaces and by the pigeon hole principle, one of them yields at most $n'/(k-1)$ points of $C$ in addition to the $\gamma(s_i), i \in [k]$.

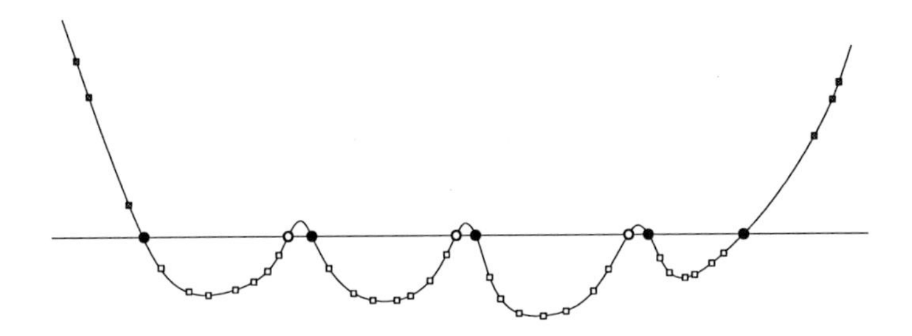

Figure 4.3: Choosing an upper half-space containing only the tails.

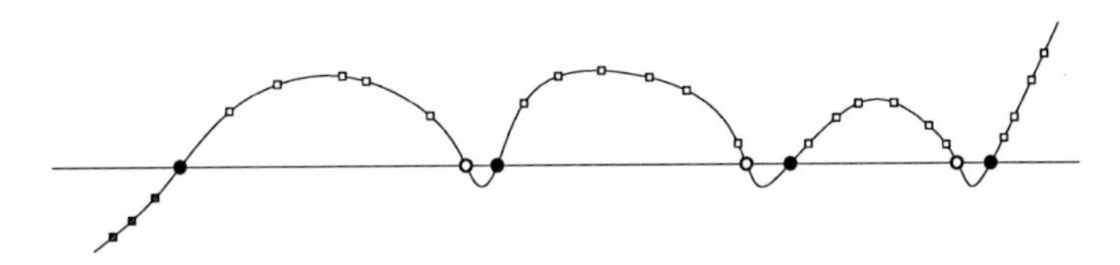

Figure 4.4: Choosing a lower half-space in odd dimension containing only the right tail.

Now we show that there is an upper half-space which contains only those points of $C$ lying to the left of $\gamma(s_1)$ or to the right of $\gamma(s_k)$. We specify the intersections to be $\gamma(s_i), i \in [k]$ and $\gamma(s_i - \delta), i \in [2, k-1]$ which yields exactly what we wanted (see Figure 4.3). This is also gives us one more choice for the general half-spaces and consequently, yields a half-space with only $n'/k$ points of $C$ in its interior.

In odd dimensions, we have an additional intersection point to specify. There is a lower half-space containing only the points of $C$ to the left of $\gamma(s_1)$: specify the intersections as $\gamma(s_i), i \in [k]$ and $\gamma(s_i - \delta), i \in [2, k]$. This has the desired properties (see Figure 4.4) and gives us one more choice which yields a lower half-space half-space with $n'/k$ points in the interior.

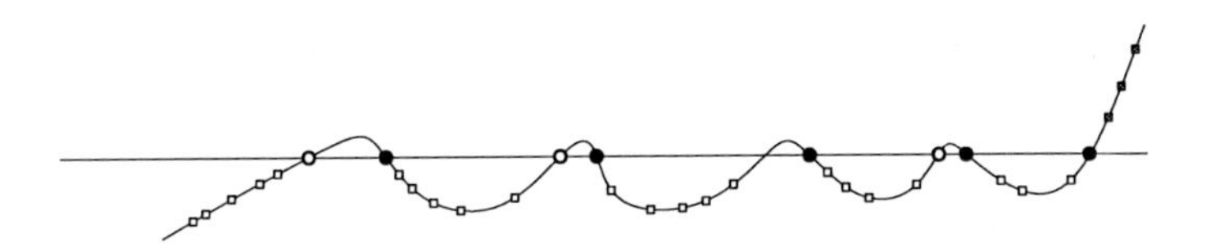

Figure 4.5: Choosing an upper half-space containing only the left tail.

Similarly, there is an upper half-space containing only those points of $C$ lying to the right of $\gamma(s_k)$: take the intersection points to be $\gamma(s_i), i \in [k]$ and $\gamma(s_i - \delta), i \in [k-1]$ (see Figure 4.5 on the previous page). This is another choice which gives a half-space with only $n'/(k+1)$ points in its interior. We can conclude:

**Lemma 4.1.** *For any $d > 2$ and $n$ sufficiently large, there is a point set $P$ in $\mathbb{R}^d$ such that for any subset $A \subseteq P$ of $k = \lfloor \frac{d}{2} \rfloor + 1$ points, there is:*

(i) *a lower half-space containing $A$ and at most $\frac{n-k}{k-1}$ other points of $P$ when $d$ is even,*
(ii) *a lower half-space containing $A$ and at most $\frac{n-k}{k}$ other points of $P$ when $d$ is odd,*
(iii) *a half-space containing $A$ and at most $\frac{n-k}{k}$ other points of $P$ when $d$ is even,*
(iv) *a half-space containing $A$ and at most $\frac{n-k}{k+1}$ other points of $P$ when $d$ is odd.*

As a consequence, we have the following upper bounds:

**Corollary 4.2.** *In odd dimensions $d$, we have*

$$\mathrm{lh}_d(n) \leq k + \frac{n-k}{k} \quad \text{and} \quad \mathrm{h}_d(n) \leq k + \frac{n-k}{k+1},$$

*whereas in even dimensions $d$ the following holds*

$$\mathrm{lh}_d(n) \leq k + \frac{n-k}{k-1} \quad \text{and} \quad \mathrm{h}_d(n) \leq k + \frac{n-k}{k}.$$

### Ball depth

For balls and points, the situation gets a bit more complicated. Essentially, the construction is again a point set on the moment curve but now we need a lemma describing intersections of the moment curve with balls (we follow the ideas from [BSSU89]).

**Lemma 4.3.** *Let $\gamma(t) = (t, t^2, \ldots, t^d)$ be the moment curve in an odd dimensional space $\mathbb{R}^d$ and $\varepsilon > 0$ sufficiently small. Consider $0 < t_1 \leq t_2 \leq \ldots \leq t_{d+1} < \varepsilon$ and denote $T := \{t_i : i \in [d+1]\}$. Furthermore, denote $T(x) := |\{i \in [d+1] : t_i < x\}|$ the number of $t_i$ smaller than $x$.*

*Then there is a ball $B(c, r)$ such that for every $x \in (0, \varepsilon)$ with $x \notin T$, the point $\gamma(x)$ lies inside $B$ if and only if $T(x)$ is odd.*

*Proof.* We need to investigate the intersections of $B(c,r)$ with $\gamma$ and find values of $c, r$ with the claimed properties. The squared distance of a given point $\gamma(t)$ on the moment curve from the centre $c$ is

$$\operatorname{dist}(\gamma(x), c)^2 = \sum_{i \in [d]} (x^i - c_i)^2.$$

The polynomial

$$\Pi(x) := \prod_{i \in [d+1]} (x - t_i)$$

vanishes on all the $t_i$ and has the sign pattern $\operatorname{sign}(\Pi(x)) = (-1)^{T(x)}$ for $x \neq t_i$ and degree $d+1$. Consider a polynomial

$$P(x) = \sum_{i=0}^{d-1} a_i x^i,$$

and assume it is non-negative on $(0, \varepsilon)$. If the $c, r$ and the polynomial $P$ satisfy the polynomial equation (both left and right hand side are polynomials in $x$)

$$L(x) := \sum_{i \in [d]} (x^i - c_i)^2 = r^2 + P(x) \cdot \Pi(x) =: R(x), \tag{4.2}$$

then we have the property that the distance of $\gamma(x)$ to $c$ is smaller than $r$ if and only if $T(x)$ is odd. Notice that $\deg(\Pi) = d+1$ and $\deg(P) = d-1$ which matches the degree $2d$ of the polynomial $L$.

Hence, it is sufficient to find the parameters, such that the equation (4.2) is satisfied (i.e. coefficients on both sides are equal) and $P > 0$ on $(0, \varepsilon)$. Keep in mind that the expressions containing a multiplicative factor of $\varepsilon$, e.g. those involving $t_i$, can be made arbitrarily small by choosing sufficiently small $\varepsilon$.

| exponent $i$ | coefficient $L(x)$ | coefficient $R(x)$ |
|---|---|---|
| $0$ | $\sum_{i \in [d]} c_i^2$ | $r^2 + O(\varepsilon)$ |
| $1 \leq i \leq d$, odd | $-2c_i$ | $O(\varepsilon)$ |
| $1 \leq i \leq d$ even | $1 - 2c_i$ | $O(\varepsilon)$ |
| $d+1 \leq i \leq 2d$ odd | $0$ | $a_{i-d-1} + O(\varepsilon)$ |
| $d+1 \leq i \leq 2d$ even | $1$ | $a_{i-d-1} + O(\varepsilon)$ |

Table 4.1: Coefficients of the polynomial equation.

The equality of the coefficients of the corresponding monomials in $L(x) = R(x)$ yields $2d+1$ equations for the $2d+1$ variables $\{a_i\}_{i=0}^{d-1}, \{c_i\}_{i=1}^{d}, r$. One can observe that this can be solved directly, since the coefficients of $x^i$ only contain the same variables as the coefficient at $x^{i+1}$ plus one additional variable. Thus, the variables can be calculated in the following order:

$$a_{d-1}, a_{d-2}, \ldots, a_0, c_d, \ldots, c_1, r$$

by a simple substitution and the equations are summarised in Table 4.1 on the preceding page.

The equations for the terms $x^i$ with $i \in [d+1, 2d]$ give

$$a_{i-d-1} = \begin{cases} 1 + O(\varepsilon) & i \text{ even}, \\ 0 + O(\varepsilon) & i \text{ odd}. \end{cases}$$

The equations for $x^i$ with $i \in [d]$ yield

$$c_i = \begin{cases} \frac{1}{2} + O(\varepsilon) & i \text{ even}, \\ O(\varepsilon) & i \text{ odd}, \end{cases}$$

and for $x^0$ we get

$$r^2 = \frac{d-1}{8} + O(\varepsilon).$$

Observe, that for sufficiently small $\varepsilon$ we have $r > 0$ and $a_0 > 0$ and also $P(x) > 0$ on $(0, \varepsilon)$. This implies the desired properties of $B(c, r)$. □

Now we know, that we can prescribe $d+1$ intersections with the moment curve in the interval $(0, \varepsilon)$ and we will always find a ball intersecting the moment curve exactly in these points. We have already $k := (d+3)/2$ points we want to have as intersections which leaves us with another $d-1-k = k-2$ free intersections points. This is the same situation as in the previous setting with the lower half-spaces in even dimension (since the moment curve "starts" and "ends" outside the balls provided by Lemma 4.3) and we can conclude:

**Lemma 4.4.** *For any $n \in \mathbb{N}$ there is a set $P$ of $n$ points in $\mathbb{R}^d$ such that for any $k := \lfloor \frac{d+3}{2} \rfloor$ points $Q \subseteq P$ there is a ball $B$ containing $k + (n-k)/(k-1)$ points of $P$ and whose boundary passes through $Q$.*

The odd-dimensional case is a consequence of the discussion above and the even-dimensional just uses the odd-dimensional construction in a hyperplane (slightly perturbed so that the point set is in general position). This implies the following bounds:

$$\mathrm{b}_d(n) \le k + \frac{n-k}{k-1} \quad \text{and} \quad \mathrm{b}_d^{\mathrm{c}}(n) \le k + \frac{n-k}{k-1}.$$

**Open problems.** A burning question for a long time has been the exact bound on the leading coefficient for the two-dimensional circle containment problem. Is $\mathrm{b}_2(n) = n/4 + o(n)$? This might be more difficult to settle than was expected and we would like to point in a slightly different direction: The presented upper and lower bounds on the problems, especially on the ball depth problem, are still far from each other in terms of the dependence on $d$. Where is the truth: is the dependence of $b_d(n)$ on $n$ and $d$ of the order $n/d$, $n/d^2$ or somewhere in between?

## 4.2 What is the point?

We showed that, in every set $P$ of $n$ points, there exists a subset $X$ of $k$ points such that $\mathrm{depth}_P^{\mathcal{H}}(X) \geq \ell$. It is natural to ask, whether it is easy to find one such point set. A trivial algorithm finds the $k$ element subset of maximum depth in time polynomial in $n$ when $d$ is fixed.

**Algorithm 4.1** MaxDepthSubset(P): finds a point set of maximum depth

$k \leftarrow \lfloor \frac{d}{2} \rfloor + 1$
$t_{OPT} \leftarrow \infty$
**for all** $X \in \binom{P}{k}$ **do**
  $t \leftarrow \mathrm{CalculateDepth}(X, P)$
  **if** $t \leq t_{OPT}$ **then**
    $t_{OPT} \leftarrow t$
    $X_{OPT} \leftarrow X$
  **end if**
**end for**
**return** $X_{OPT}, t_{OPT}$

This algorithm performs $O(n^{\lfloor d/2 \rfloor + 1})$ depth calculations. Our proof of Theorem 3.2, however, yields a faster algorithm which finds a set $X$ of depth at least $(1-\varepsilon)\ell + O(1)$ using only few depth calculations: Recall that in the proof of Theorem 3.2 we used an artificial constant $\alpha$ (which was equal to one for our purposes) in (3.15). This constant controls what fraction of all subsets $X$ of cardinality $k$ we allow to have depth at most $\ell$. We obtained the largest $\ell$ for which our method guarantees this condition as a function in $\alpha$ (we will write $\ell(\alpha)$ to stress this). By (3.2.2) the ratio $\ell(1-\varepsilon)/\ell(1)$ is

$$\frac{\ell(1-\varepsilon)}{\ell(1)} = \frac{\frac{1-\varepsilon}{2_d c_{\mathcal{R}} e + 1 - \varepsilon}}{\frac{1}{2_d c_{\mathcal{R}} e + 1}} + o_n(1) \geq \frac{\frac{1-\varepsilon}{2_d c_{\mathcal{R}} e + 1}}{\frac{1}{2_d c_{\mathcal{R}} e + 1}} + o_n(1) = 1 - \varepsilon + o_n(1).$$

This tells us that if we want an $\varepsilon$ fraction of all the $\binom{n}{k}$ possible sets $X$ to have depth at least $\ell'$, we only need to sacrifice an $\varepsilon$ fraction of $\ell$. Hence, a random set $X$ of cardinality $k$ has a probability $\varepsilon$ of having depth at least $\ell(1-\varepsilon)$. Out of $1/\varepsilon$ such random sets $X$, at least one has depth more than $\ell(1-\varepsilon)$ with probability at least $(1-\varepsilon)^{1/\varepsilon} \approx 1/e$ for $\varepsilon$ small. This motivates Algorithm 4.2.

**Algorithm 4.2** ApproxDeepSubset(P): finds a point of depth at least $(1-\varepsilon)\cdot\ell$

$k \leftarrow \lfloor \frac{d}{2} \rfloor + 1$
**for** $i := 1$ to $1/\varepsilon$ **do**
  $X \leftarrow \binom{P}{k}$ uniformly at random
  $t \leftarrow$ CalculateDepth$(X, P)$
  **if** $t \geq (1-\varepsilon)\cdot\ell$ **then**
    **return** $X$
  **end if**
**end for**
**return** failed

This algorithm outputs a subset $X$ of desired depth with probability $1/e$ and fails otherwise. Sufficiently many repetitions make the failure probability arbitrarily small.

The only remaining ingredient is the actual calculation of the depth (or testing whether it is large enough). It is sufficient to consider all the half-spaces defined by $d$ points out of which at least one is in the set $X$: The reason is that every other half-space can be first translated (without gaining any new points) until it contains a point of $X$ on its boundary. One can then rotate it in such a way that it contains exactly $d-1$ other points on its boundary while gaining at most these $d-1 = O(1)$ points. Moreover, there is an arbitrarily small perturbation of the new half-space which contains the same number of points as the original one.

Hence, the following simple brute-force algorithm works: iterate over all the subsets $A$ of points containing one point of $X$ and $d-2$ points of $P$ (possibly also in $X$). This still gives us one degree of freedom to rotate the half-space around the aff$(A)$. Sort the remaining points angularly around aff$(A)$. By a single linear pass, it is now possible to calculate the number of points in each of the half-spaces defined by $A$ and one other point (it is also necessary to test, whether it contains $X$ but that can be done in $O(2^d) = O(1)$ time since $d$ is constant). For a precise calculation, the possible $O(2^d)$ combinatorially different perturbations have to be checked as well. The algorithm needs to go over $O(n^{d-2})$ sets $A$ and, for each one of them, perform $O(n \log n)$ operations. This gives a total running time of $O(n^{d-1} \log n)$.

For the two algorithms for finding the centre points, this yields a running time of $O(n^{d+\lfloor d/2 \rfloor} \log n)$ for the exact procedure and $O(\varepsilon^{-1} n^{d-1} \log n)$ for the randomised approximate algorithm with constant success probability.

**Open problem.** We did not particularly try to search for a fast algorithm but rather wanted to show the consequences of our proof. Improving its running time might be worth some effort.

# Part II

# Crossings

# 5

# $k$-edges and crossings: background

The main theme in Part II are $k$-facet crossings in higher dimension. We examine two major directions: a $k$-facet crossing identity in $\mathbb{R}^3$ and bounds on a higher-dimensional generalisation of the rectilinear crossing number of complete graphs.

The notion of $k$-sets has been around for decades and crossings of $k$-edges and $k$-facets have successfully been used to prove lower bounds on the number of $k$-sets in low dimensions. A particularly inspiring result was found by Andrzejak et al. [AAHP$^+$98]. The authors proved an identity relating the number of crossings between $k$-edges in the plane to some better understood quantities. Their result gives an alternative proof of the upper bound on the number of $k$-edge crossings which is the main ingredient of Dey's upper bound on the number of planar $k$-sets [Dey98].

Analogously to the plane, a crossing identity of a similar flavour in higher dimensions could help us understand $k$-facets and their crossings better. Ultimately, it might contribute to an improvement of the upper bound on the number of $k$-sets. We made first steps in that direction and hope they are going to bear fruits in the future.

The rectilinear crossing number of complete graphs has been studied extensively and, only recently, it was shown that it differs in the asymptotically leading term from the usual crossing number. The authors, Lovász et al. [LVWW04], show this by improving lower bounds on the rectilinear crossing number. A similar result was discovered independently by Ábrego and Fernández-Merchant [AFM05] but the authors of the latter did not aim for the additional $\varepsilon$ needed to show the difference in the leading terms. Further improvements of the lower bounds were found in the consequent works of Balogh and Salazar [BS04, BS06], Aichholzer et al. [AGOR07] or the work of Ábrego et al. [ÁBFM$^+$08], to name a few (a more exhaustive list of references is provided in the pages to follow). A possible interpretation of the rectilinear crossing number of $K_n$ is as the minimum number of convex quadrilaterals in a set of $n$ points. We are interested in the potential generalisations to higher dimensions and give non-trivial lower bounds for one of the variants.

This chapter should serve as an introduction, where we present to the reader the necessary background to the problems we want to tackle and we introduce all the notions needed to state and understand our results. We start with a brief detour to the results about $k$-sets and afterwards we give a short overview of rectilinear crossing numbers of complete graphs. The proof of our crossing identity is a little demanding in terms of the number of different notions it uses. These notions are introduced here and, among others, include $k$-facets, spherical $k$-facets, a three-dimensional notion of crossings, contractions and $g$-vectors of point configurations. We conclude this introductory chapter by showing the basic ingredients of the planar crossing identity of Andrzejak et al. [AAHP$^+$98].

In the following chapters, we present our results: In Chapter 6, we prove a crossing identity on the 2-sphere by a continuous-motion argument. First, we concentrate on the specifics of continuous motion in three dimensions and on the 2-sphere. We continue by applying this and proving a spherical generalisation of Lovász lemma and the spherical crossing identity. We finish the chapter by discussing implications of this identity. In Chapter 7, we discuss two possible generalisations of the rectilinear crossing number of a complete graph to higher dimensions. There are at least two reasonable generalisations and for one of them, we show non-trivial lower and upper bounds inspired by the approach of Lovász et al. [LVWW04] in the plane.

## 5.1 $k$-edges and crossings

Given a set $P$ of $n$ points in general position in $\mathbb{R}^d$ and an integer parameter $k$, a *$k$-set* of $P$ is a subset $S \subseteq P$ of size $|S| = k$ that can be strictly separated

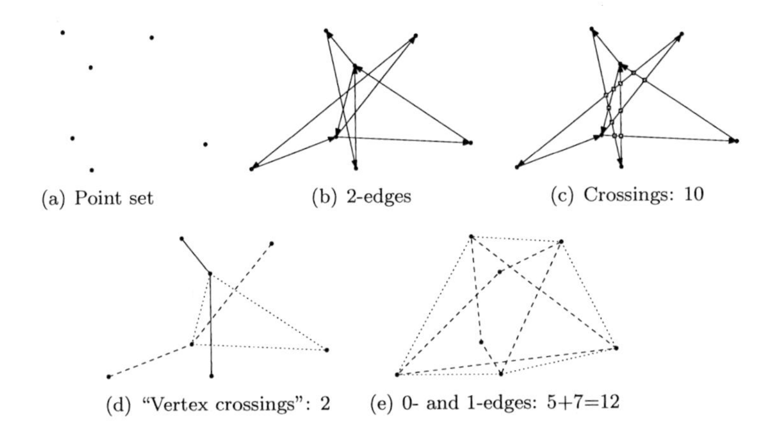

(a) Point set (b) 2-edges (c) Crossings: 10

(d) "Vertex crossings": 2 (e) 0- and 1-edges: 5+7=12

Figure 5.1: Example of the identity for 2-edges

from its complement $P \setminus S$ by an affine hyperplane. The *$k$-set problem* is: What is the maximum number of $k$-sets of an $n$-element point set in $\mathbb{R}^d$? This has been a basic open problem in discrete and computational geometry for more than thirty years[*] and despite intensive research, there still remain substantial gaps between the known upper and lower bounds, even in low dimensions.

In the plane $\mathbb{R}^2$, the currently best upper bound $O(n\sqrt[3]{k})$ is due to Dey [Dey98]. Like most papers dealing with $k$-sets, Dey does not work directly with $k$-sets but with the equivalent (up to constant factors) notion of *$k$-edges*, i.e. directed edges $pq$ spanned by points $p, q \in P$, such that there are exactly $k$ points of $P$ to the right of (the line through) the segment $pq$. A key ingredient in Dey's proof is to show that the number of crossings between $k$-edges is at most $O(n(k+1))$. Dey's analysis was further refined by Andrzejak et al. [AAHP+98] who proved the following *crossing identity* for $k$-edges:

$$\operatorname{cr}_k(P) + \sum_{q \in P} \binom{\deg_k(q)}{2} = e_{<k}(P), \tag{5.1}$$

where $k < \frac{n-2}{2}$ (for $k = \frac{n-2}{2}$ the formula is slightly different). Here, $\operatorname{cr}_k$ is the number of crossings of $k$-edges, $\deg_k(q)$ is *half* of the number of $k$-edges emanating from[ℸ] $q$ and $e_{<k}$ is the total number of $j$-edges, $0 \le j < k$. This

[*]Some of the key references are [Lov72, ELSS73, GP84, EW85, PSS92, BFL90, ŽV92, ABFK92, DE94, Dey98, AAHP+98, SST01, Tót00, MSSW06]; for further references and background, we refer to the survey [Wag08] or to [Mat02, Chapter 11].

[ℸ]The number of these edges is always even. If one takes the orientation of the $k$-edges identity implies the desired bound on $\mathrm{cr}_k$, since it is known [AG86, Pec85] that $e_{<k} \leq nk$.

**Example 5.1.** *To get some feeling, let us look at an example in Figure 5.1 on the previous page. We consider the identity for* $k = 2$. *We find* 10 *proper crossings of the* 2*-edges. Concerning the second term, there are only two points of degree* 2 *and all the other points have degree* 1 *and hence, the second term is* $1 + 1 = 2$. *On the right hand side, we have* 5 0*-edges plus* 7 1*-edges which makes* 12 $(< 2)$*-edges altogether.*

In dimension 3, one considers $k$-facets: triangles spanned by three points of $P$ with precisely $k$ points on a specified side. The best upper bound $O(nk^{3/2})$, due to Sharir et al. [SST01], is also proved by analysing certain crossing configurations between $k$-facets, namely *pinched crossings* (two triangles that share a vertex and intersect in their relative interiors; see Figure 5.5 on page 70), but an exact identity like (5.1) was still missing.

Here, we prove such an identity which we believe adds to our understanding of $k$-facets in $\mathbb{R}^3$. As a key technical step along the way, we first extend (5.1) to point sets on the 2-sphere. As a new ingredient (compared to the planar case), our identities involve the winding number of $k$-facets around a given point in 3-space, as introduced by Lee [Lee91] and by Welzl [Wel01].

## 5.2 Rectilinear crossing number

Given a graph $G$, the minimum number of edge crossings in a straight-line drawing of $G$ in the plane is called *rectilinear crossing number* of $G$ and denoted by $\overline{\mathrm{cr}}(G)$. The rectilinear crossing number $\overline{\mathrm{cr}}(K_n)$ of a complete graph is the minimum number of convex quadrilaterals in any $n$-point set (since endpoints of two intersecting edges form a convex quadrilateral and vice versa, a convex quadrilateral induces a crossing of its diagonals):

$$\overline{\mathrm{cr}}(K_n) = \square(n) := \min_{S \subseteq \mathbb{R}^2, |S|=n} \square(S),$$

where $\square(S)$ denotes the number of convex quadrilaterals in a point set $S$.

A breakthrough was made by Lovász et al. [LVWW04] and independently by Ábrego and Fernández-Merchant [AFM05] who related the rectilinear crossing number to the numbers $e_{\leq k}$ of $(\leq k)$-edges in the plane. They gave a lower

into account, it is possible to talk about the indegree or the outdegree instead to avoid the half. The indegree is equal to the outdegree as a consequence of the interleaving property (Fact 5.2) which we will encounter soon.

bound for $e_{\leq k}$ and, as a consequence, obtained a considerably improved lower bound for the rectilinear crossing number of $K_n$. The former paper added a further small improvement which implies that the rectilinear crossing number of $K_n$ differs in the asymptotically dominant term from the usual crossing number:

$$\square(P) \geq \left(\frac{3}{8} + \varepsilon\right)\binom{n}{4}.$$

Later papers [BS04, BS06, AGOR07, ÁBFM+08, ÁFMLS08, ÁCFM+11] gave improvements by fighting for better lower bounds on the numbers $e_{\leq k}$ for certain ranges of $k$:

$$0.380488\binom{n}{4} \geq \square(n) \geq 0.37997\binom{n}{4}.$$

The upper bound follows from the construction of [ÁFM07] using an initial configuration from [ÁCFM+10].

# 5.3 Basics and preliminaries

Let us introduce notation and terminology for the rest of Part II. We will see $k$-edges and $k$-facets in the space as well as on the sphere, three-dimensional crossings, contractions and $f, g$ and $h$-vectors.

## Points and vectors

In the rest of the text we will clearly differentiate between *point sets* with no distinguished origin, for which we are only interested in affine properties, and *vector configurations* with a distinguished origin $\mathbf{0}$ where linear properties involving this origin come into play as well.

## $k$-facets

Let $P$ be a set of $n$ points in $\mathbb{R}^d$ in *general position*, i.e. any $d+1$ or fewer points are affinely independent. Consider a $(d-1)$-dimensional simplex $\sigma := \sigma(p_1, \ldots, p_d) := \operatorname{conv}\{p_1, \ldots, p_d\}$ spanned by $d$ distinct points $p_1, \ldots, p_d \in P$. We will identify the simplex$\sigma$ with the set of points spanning it and write $\sigma = p_1 \ldots p_d$. The affine hull of $\sigma$ is a hyperplane which divides $\mathbb{R}^d$ into two open half-spaces. A *coorientation* of the simplex $\sigma$ is declaring one of these two half-spaces positive and the other negative, denoted by $\sigma^+$ and $\sigma^-$, respectively. A cooriented simplex is called a *$k$-facet* if it contains exactly

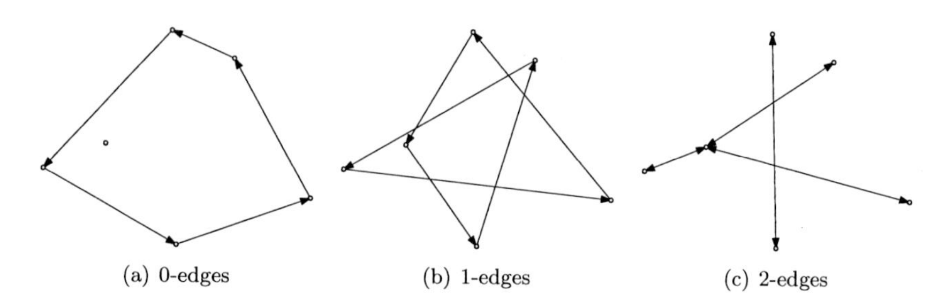


Figure 5.2: $k$-edges

$k$ points of $P$ in its positive half-space. In particular, the 0-facets of $P$ are precisely the facets of its convex hull with coorientation given by the outer normal vectors. We denote the number of $k$-facets of $P$ by $e_k(P)$, or simply $e_k$ if $P$ is clear from the context. Note that $e_k = 0$ for $k < 0$ or $k > n - d$ and that $e_k = e_{n-d-k}$ for all $k$ by simply reversing the coorientations. The number of ($\leq k$)-facets and ($< k$)-facets of $P$ will be denoted by $e_{\leq k}(P)$ and $e_{<k}(P)$, respectively (i.e. $e_{\leq k}(P) := \sum_{i=0}^{k} e_i(P)$ and $e_{<k}(P) := \sum_{i=0}^{k-1} e_i(P)$) or only by $e_{\leq k}$ and $e_{<k}$ if the point set $P$ is understood from the context.

## Spherical $k$-facets

The $d$-dimensional unit sphere in $\mathbb{R}^{d+1}$ centred at the origin $\mathbf{0}$ will be denoted by $\mathbb{S}^d$ (for example, $\mathbb{S}^1$ is the unit circle in the plane). Let $V$ be a set of $n$ vectors on[⁊] $\mathbb{S}^d$ in *linearly general position* (i.e. any $d+1$ or fewer points linearly independent). Given a set of $d$ vectors $v_1, \ldots, v_d \in V$, we will call the set $\sigma := \sigma(v_1, \ldots, v_d) := \mathbb{S}^d \cap \operatorname{cone}\{v_1, \ldots, v_d\}$ a *spherical simplex* spanned by those vectors (see Figure 5.3 on the next page for an example of a spherical simplex on the 2-sphere). There is a unique $(d-1)$-dimensional unit sphere $S$ centred at $\mathbf{0}$ containing $\sigma$, and it subdivides $\mathbb{S}^d$ into two open hemispheres. Any such sphere will be called a *great circle* of $\mathbb{S}^d$. A coorientation of $\sigma$ is again declaring one of the hemispheres positive and the other negative, denoted by $\sigma^+$ and $\sigma^-$, respectively. A cooriented spherical simplex is called a *(linear) k-facet*, if it contains exactly $k$ points of $V$ in its positive hemisphere $\sigma^+$. A spherical $k$-facet $v_1 \ldots v_d$ defines a $k$-facet $v_1 \ldots v_d\mathbf{0}$ (with consistent coorientation) incident to $\mathbf{0}$ in the point set $V \cup \{\mathbf{0}\} \subseteq \mathbb{R}^{d+1}$ and vice versa.

[⁊]Technically, $\mathbb{S}^d$ is not a vector space and it might sound strange to talk about vectors on the sphere. Nevertheless, we will only use properties depending on the direction but not the length and thus, one can imagine non-zero vectors in $\mathbb{R}^{d+1}$ instead.

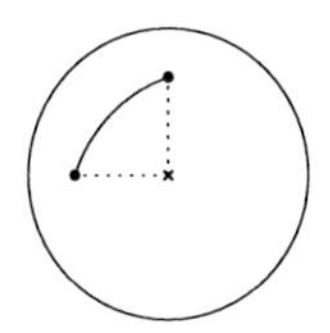

Figure 5.3: Spherical edge

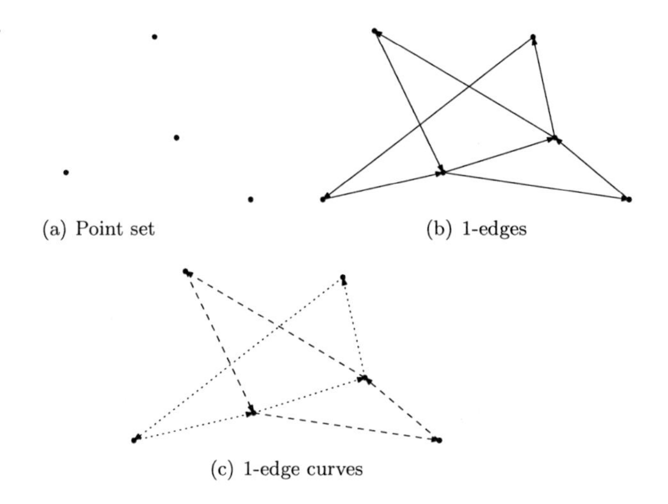

(a) Point set

(b) 1-edges

(c) 1-edge curves

Figure 5.4: A $k$-edge curve in the plane

Planar $k$-facets and linear $k$-facets on the two-dimensional sphere are called *$k$-edges* and an orientation of a $k$-edge by convention uniquely defines its coorientation by declaring the right half-plane to be positive and vice versa, a coorientation defines the orientation. In the figures, we will implicitly give the coorientation by orienting the edges (as we do in Figure 5.2 on the facing page). Similarly, $k$-facets in $\mathbb{R}^3$ and linear $k$-facets on $\mathbb{S}^3$ are called *$k$-triangles.*

## Crossings

Let $P$ be a set of $n$ points in $\mathbb{R}^d$ (or vectors in $\mathbb{S}^d$), $k < \frac{n-d}{2}$ and $q_1, \ldots, q_{d-1} \in P$. The degree $\deg_k(q_1 \ldots q_{d-1})$ denotes *half* the number of $k$-facets of $P$ defined by $q_1, \ldots, q_{d-1}$ and one additional point of $P$.

The half comes from an interpretation of the $k$-facets as a "hypersurface".

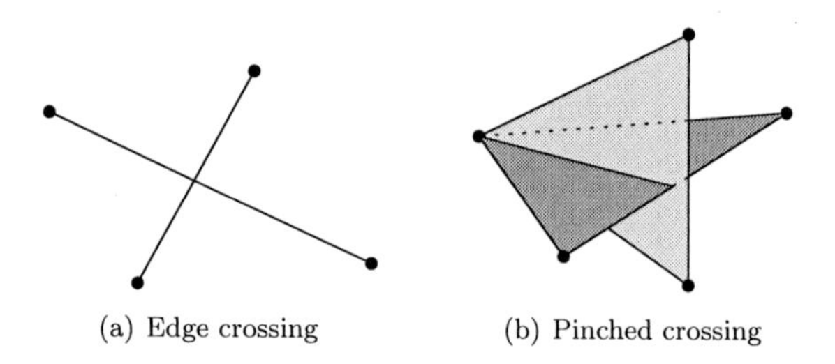

(a) Edge crossing (b) Pinched crossing

Figure 5.5: Crossing in the plane and in the 3-space.

Let us explain what we mean by that: In $\mathbb{R}^2$ this means a union of closed curves (non-smooth and typically heavily self-intersecting) and in a higher dimension a generalisation thereof. In our setting with $(d-1)$-dimensional simplices ($k$-facets) a "continuation" of every simplex over each of its $(d-2)$-dimensional faces has to be well defined.

For a $k$-facet $\sigma = \sigma(q_1, q_2, \ldots, q_d)$ and its $d-1$ points $A = \{q_1, \ldots, q_{d-1}\}$ one can define a neighbouring $k$-facet over $A$ as the first $k$-facet encountered by rotating $\sigma$ around $A$ in the negative direction (i.e. the remaining point $q_d$ becomes part of the negative half-space $\sigma^-$). This neighbourhood relation is well defined due to the interleaving property of $k$-facets (we will see it in Fact 5.2) and hence, one can view the $k$-facets as a (locally defined) $(d-1)$-dimensional hypersurface[§] $\Sigma_k$. The degree $\deg_k(q_1 \ldots q_{d-1})$ is then just the number of *sheets* of the $k$-facet hypersurface passing through $A$ – that is, the number of pairs of neighbouring (in the sense described above) $k$-facets passing through $q_1 \ldots q_{d-1}$. For an example of the $k$-edge curve in the plane, consult Figure 5.4 on the previous page.

Two *$k$-facets $\sigma$ and $\tau$ cross*, if their relative interiors intersect[¶] (denoted by $\text{relint}(\sigma) \cap \text{relint}(\tau) \neq \emptyset$) and a *line $\ell$ crosses a $k$-facet $\sigma$*, if $\ell \cap \text{relint}(\sigma) \neq \emptyset$.

Let $P$ be a set of $n$ points in $\mathbb{R}^2$ or vectors on $\mathbb{S}^2$. The number of pairs of crossings of $k$-edges in $P$ will be denoted by $\text{cr}_k(P) := |\{\{pq, rs\} \mid pq \neq rs \text{ are crossing } k\text{-edges}\}|$. A *generalised crossing* is either a proper crossing of $k$-edges or a *degenerate crossing*, i.e. a pair of branches of the $k$-edge curve passing through a common vertex.

Let $P$ be a set of $n$ points in $\mathbb{R}^3$ or vectors on $\mathbb{S}^3$ and let $opq, ors$ be two

[§] Notice that, although the transition over every $d-1$ dimensional feature is well defined, there might be many degeneracies of the hypersurface on features of lower dimension, such as very wild neighbourhoods of a vertex etc.

[¶] The $k$-facets are $(d-1)$-dimensional objects in $\mathbb{R}^d$, therefore they have an empty interior. Thus we need to speak of their relative interiors instead.

distinct $k$-facets of $P$ sharing the vertex $o$. We say that $opq$ and $ors$ form a *pinched crossing* if their relative interiors intersect ($\text{relint}(opq) \cap \text{relint}(ors) \neq \emptyset$; see Figure 5.5 on the facing page). The number of pairs of $k$-facets of $P$ forming a pinched crossing will be denoted by $\text{pcr}_k(P)$.

## Contractions

Let $P$ be a set of points in $\mathbb{R}^d$ in general position and $S \subset P, |S| < d$. Let $P'$ denote a point set obtained by projecting $P \setminus S$ orthogonally onto $\text{aff}(S)^{\perp}$ (the affine orthogonal complement of the affine hull of $S$). Set the coordinate system such that the projection of $S$ is $\mathbf{0}$. Now projecting $P'$ centrally onto a $(d-|S|)$-dimensional sphere yields a vector set on $\mathbb{S}^{d-|S|}$, which is the *contraction* $P/S$. The image of a point $p \in P$ or a set $Q \subseteq P$ under this mapping will be denoted by $p/S$ or $Q/S$ respectively (assuming $P$ is understood from the context).

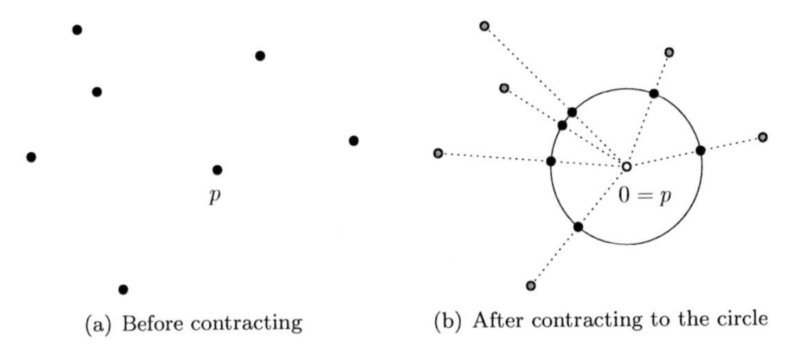


(a) Before contracting (b) After contracting to the circle

Figure 5.6: A contraction of a planar point set

Similarly, by taking the (linear) orthogonal complement of the (linear) span, we define contraction of a vector configuration.

Observe that the contraction $U/S$ of a $k$-facet $U \cup S$ of $P$ is again a $k$-facet in $P/S$ and vice versa, if $V$ is a $k$-facet of $P/S$ then its preimage together with $S$ is a $k$-facet of $P$ incident to $S$. Thus, $k$-facets of $P/S$ are in one-to-one correspondence with $k$-facets of $P$ containing $S$. For $q \in P$, the contraction $P/p := P/\{p\}$ is just the “angular view” of $p$ onto $P \setminus \{p\}$ (see Figure 5.6). When $P \subset \mathbb{R}^3$ and $q \in P$, one can see that $k$-facets in $P$ containing $q$ and $k$-edges in $P/q$ have the same “shape”, i.e. two $k$-facets $U, V \ni q$ of $P$ form a pinched crossing (centred at $q$) if and only if their contractions, $k$-edges $U/q$ and $V/q$, cross. If $q$ is a convex hull vertex of $P$, the contraction $P/q$ lies completely in one hemisphere of $S^2$ and thus, can be treated as a planar point set (for the purposes of $k$-edges and their crossings – as we can centrally project it onto a plane close to $q$ with the same normal vector as the aforementioned hemisphere; see Lemma 6.4 on page 81 for a detailed statement).

## $f$-, $g$- and $h$-vectors

Let $V$ be a set of vectors in general position in $\mathbb{R}^d$. For $0 \leq k \leq n-d-1$, we define $f_k$ as the number of subsets $U$ of size $d+k+1$ which contain $\mathbf{0}$ in the interior[‖] of their convex hull $\operatorname{conv}(U)$ (or alternatively, $\mathbb{R}^d = \operatorname{cone}(U)$).

$$f_k = f_k(V) := |\{U \subset V : |U| = d+k+1, \mathbf{0} \in \operatorname{int}(\operatorname{conv}(U))\}|.$$

The integer vector $(f_0, f_1, \ldots, f_{n-d-1})$ is called the *$f$-vector* of $V$. It is closely related to the *$h$-vector* $(h_0, h_1, \ldots, h_{n-d-1})$ which is defined by inverting the system of equations[**]

$$f_k = \sum_j \binom{j}{k} h_j, \qquad 0 \leq k \leq n-d-1.$$

It is convenient to extend the range of indices to all integers by defining $f_k := h_k := 0$ for $k < 0$ or $k > n-d-1$. Furthermore, we define $g_k := h_k - h_{k-1}$. We collect the terms $g_k$ into the *$g$-vector* $(g_0, g_1, \ldots, g_{n-d-1}, g_{n-d})$ (note that the range of indices is larger by 1 than that of the $f$-vector and the $h$-vector). These definitions are *Gale dual* to the more common definition of $f$, $h$, and $g$-vectors for simplicial polytopes, see [Wel01].

**Winding numbers.** Lee [Lee91] and Welzl [Wel01] independently observed the following geometric interpretation of the numbers $g_k$ as winding numbers. Consider $U \subset \mathbb{R}^d$, which we think of as an affine point set. Consider the (affine) $k$-facets of $U$. Let $o$ be a point that does not lie on any of the $k$-facets of $U$ and let $\rho$ be a semi-infinite ray directed towards $o$ that avoids any $(d-2)$-dimensional affine flat spanned by $U$. Let $g_k^+(U, \rho, o)$ be the number of $k$-facets that we *enter* (traverse from the negative to the positive side) as we move from infinity towards $o$ along $\rho$, and let $g_k^-(U, \rho, o)$ be the number of $k$-facets of $U$ that we *leave* (traverse from the positive to the negative side). It turns out that the difference $g_k^+(U, \rho, o) - g_k^-(U, \rho, o)$ is independent of the ray $\rho$, and coincides with $g_k(U)$, if we consider $U$ as a vector configuration translated by $-o$ (thus, taking $o$ as the origin $\mathbf{0}$). See Figure 5.7 for an illustration.

[‖]The interior is taken with respect to the standard topology of $\mathbb{R}^d$, i.e. the convex hull has to be full-dimensional in order to have a non-empty interior.

[**]In terms of generating functions, if we set $f(x) := \sum_k f_k x^k$ and $h(x) := \sum_k h_k x^k$, the equations yield $f(x) = h(x+1)$, i.e. $h(x) = f(x-1)$, i.e. $h_j = \sum_k (-1)^k \binom{k}{j} f_k$.

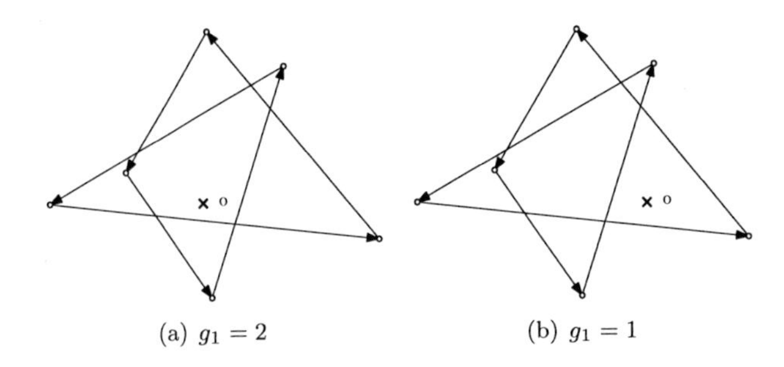


(a) $g_1 = 2$ (b) $g_1 = 1$

Figure 5.7: Examples of $g$-values

## 5.4 Planar crossing identity

We now review some of the basic ingredients of the proof of the crossing identity (5.1). Throughout this section, let $P$ be a set of $n$ points in the plane.

One of the most basic, nevertheless very useful, structural results about $k$-edges is the following property.

**Fact 5.2** (Interleaving property). *Let $p \in P$. For every pair of $k$-edges $ap, bp$ directed towards $p$ there is a $k$-edge $pc$ directed away from $p$ such that $p$ lies in the triangle $abc$ (in another words, $c$ lies in the angle opposite to the angle $apb$; see Figure 5.8). Similarly, for every two outgoing $k$-edges there is some incoming edge in the opposite angle.*

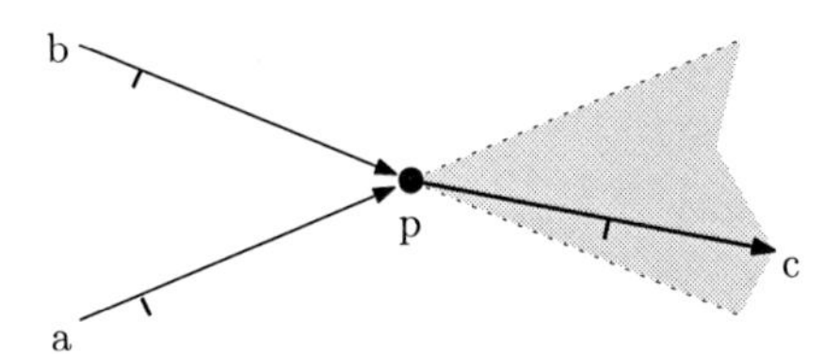


Figure 5.8: Interleaving property for the $k$-edges

In fact, one can analyse the arrangement of $k$-edges around a point even more precisely.

**Fact 5.3.** *Let $k < \frac{n}{2}$ and $\ell$ be a line passing through a point $p \in P$ but no other point of $P$ and let $\ell^+$, $\ell^-$ be the half-planes defined by $\ell$. Denote by $E_k(p)$ the set of all $k$-edges incident to $p$, and let $E_k^+(p)$ and $E_k^-(p)$ denote the* *sets of those k-edges incident to p whose remaining endpoint lies in* $\ell^+$ *and* $\ell^-$*, respectively. Then the difference* $|E_k^+(p)| - |E_k^-(p)|$ *is (see Figure 5.9):*

$$\begin{array}{ll} +2 & \text{if } |(P \setminus p) \cap \ell^+| \leq k \\ -2 & \text{if } |(P \setminus p) \cap \ell^-| \leq k \\ 0 & \text{otherwise} \end{array}$$

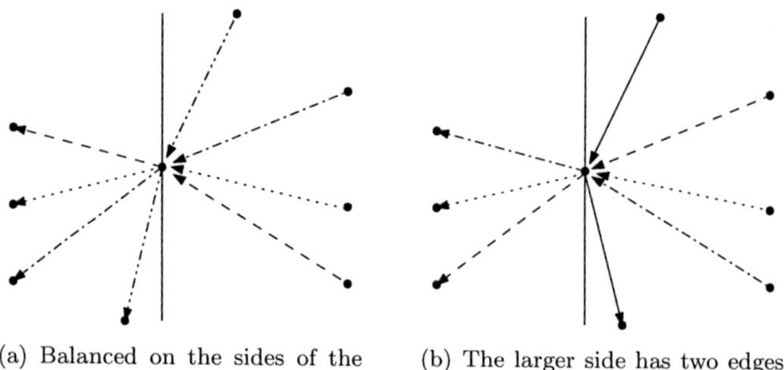

(a) Balanced on the sides of the line

(b) The larger side has two edges more

Figure 5.9: $k$-edges around a point

This directly implies[JJ] a 2-dimensional exact variant of the so-called *Lovász lemma*, which is one of the basic tools to prove bounds on $k$-sets, both in the plane and in higher dimensions.

**Fact 5.4** (Lovász lemma [Lov72]). *Let* $\ell$ *be a line not passing through any point of* $P$ *and* $\ell^+$ *and* $\ell^-$ *the half-planes defined by it. Then* $\ell$ *intersects exactly* $e_k(P, \ell) := 2 \cdot \min\{k, |P \cap \ell^+|, |P \cap \ell^-|\}$ *k-edges of* $P$.

Andrzejak et al. [AAHP+98] proved the crossing identity (5.1) by analysing how the involved quantities change under continuous motion of the point set, and the facts collected above are the basic ingredients that allow one to perform this analysis.

[JJ]It follows from a one-dimensional continuous motion argument. To sketch it, consider a line $\ell_0$ parallel to $\ell$ disjoint from the convex hull of $P$. This line does not intersect any $k$-edge of $P$. Translating $\ell_0$ into $\ell$ changes the number of intersected $k$-edges whenever the line sweeps over a vertex and the change is exactly described by Fact 5.3.

# 6

# Crossing identity on the sphere

After the previous introductory chapter, we are ready to start proving theorems. In this chapter, we generalise the crossing identity (5.1) proved by Andrzejak et al. [AAHP$^+$98] to the sphere and study the consequences of this generalisation. As for the aforementioned identity, we give a proof based on a continuous motion argument — although our argument gets slightly more complicated.

In Section 6.1, we discuss the basic idea of continuous motion proofs. Then, we specifically look at continuous motion in $\mathbb{R}^3$ and on the sphere $\mathbb{S}^2$, where we study the effect of the motion on $k$-facets, their crossings and the winding numbers $g_k$. We classify the events when some of these quantities change and give illustrations of what happens during these events.

In Section 6.2, we prove a generalisation of the Lovász lemma (Fact 5.4) to the sphere $\mathbb{S}^2$ (in Theorem 6.6). This is a key lemma for our further progress and allows us to generalise the identity (5.1) to the sphere (Theorem 6.7). As a consequence of this identity we also obtain a weaker identity in the space $\mathbb{R}^3$ (Corollary 6.8) and a generalisation of Dey's upper bound on the number of $k$-sets on the sphere $\mathbb{S}^2$ (Corollary 6.10).

In Section 6.3, we discuss what consequences our theorems do or do not have and why.

This is joint work with Uli Wagner [SW10].

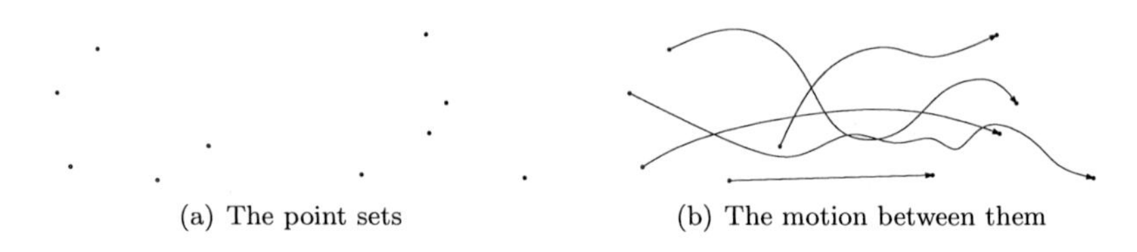

(a) The point sets (b) The motion between them

Figure 6.1: An example of continuous motion

## 6.1 Continuous motion

As we have already mentioned, our proofs in this chapter are based on continuous motion arguments. In this section, we explain the basic idea of continuous motion proofs and then go deeper into the specifics in $\mathbb{R}^3$ and $\mathbb{S}^2$ for proving theorems about $k$-edges and $k$-facets. We introduce a notion of transition — an event when something important happens — and classify the transitions in $\mathbb{R}^3$ and $\mathbb{S}^2$. At the end of the section, we discuss the changes of several important quantities during the transitions.

### 6.1.1 Basics of continuous motion

We start with the idea of continuous motion proofs. Imagine you want to show an equality of some quantities for all point sets in the plane, say. It is typically easy to show that the equality holds for some particular point set (convex position would be a usual suspect for many questions). If the quantities we are interested in behave in some sense locally, one can just try to move the points from the point set that has already been analysed into any particular point configuration and check the equality throughout. The changes in the quantities often happen on some very special discrete events and it is sufficient to have control over them.

Let us now start formally and look at $k$-facets of the point set. Although we only introduce the notation for point sets, we will use analogous terminology for vector configurations. Let $P(0) := P$ be a set of $n$ points $\{p_1, \dots, p_n\}$. We consider the changes in $k$-facets under a *generic continuous motion* $P(t) = \{p_1(t), \dots, p_n(t)\}$. This means that each $p_i(t)$ depends continuously on time $t$, and that $P(t)$ is in general position throughout, except for a finite number of *critical instants* $t_1, t_2, t_3, \dots$ at which only one (proper) degeneracy happens each (for example in $\mathbb{R}^3$ exactly one quadruple of points becomes coplanar but no three points are collinear; on $\mathbb{S}^2$ exactly one triple lies on a common great circle; furthermore, we usually insist that the combinatorial type of the points involved is different before and after the degeneracy). These events of degeneracy will be called *transitions* (the exact definition of a degeneracy is problem specific: these are the situations when something important happens).

We are only interested in $\mathbb{S}^2$ and the further discussion will be specific to $\mathbb{S}^2$ and $\mathbb{R}^3$ (our terminology originates from $\mathbb{R}^3$ and therefore, it is more natural to present it in $\mathbb{R}^3$ first). Let us start in the space $\mathbb{R}^3$. Consider a set $P \subset \mathbb{R}^3$ of $n$ points. A *transition* is a coplanarity of four distinct points on a common plane $h$ such that one of the points changes the side of the plane (considering the plane as defined by the other three points). We will observe (in Observation 6.1) that this is the only way, in which the structure of the $k$-triangles of $P$ can change. The four points involved in the transition either form a convex quadrilateral at the critical instant and we speak of a *convex transition* or the points form a triangle with a point inside and we speak of a *non-convex transition* or a *transition through a triangle.* The *order* of a transition is the number $|P \cap h^+|$ at the critical instant.

For a configuration $V \subset \mathbb{S}^2$ of $n$ vectors, a transition is a triple of vectors lying on a common great circle (one of them changing side) or in other words, a transition is a coplanarity of a triple of points of $V$ and $\mathbf{0}$. This point of view allows us to adopt the same terminology as in $\mathbb{R}^3$.

### 6.1.2 Relevant events

This subsection will clarify which transitions on $\mathbb{S}^2$ are relevant for us. We want to know what effect continuous motion has on the following objects:

- spherical $k$-edges,
- crossings between them, and
- the winding numbers $g_i$.

The combinatorial structure of the $k$-edges, that is the set of pairs of points forming a $k$-edge, only changes when some $k$-edge appears or disappears. It is sufficient to look at the events when a $k$-edge disappears, since an appearance of a $k$-edge can be viewed as time reversal of a disappearance. A $k$-edge disappears when a point travels from of its sides to the other, i.e. during a transition of order $k$ (a point entering its positive side) or a transition of order $k-1$ (a point leaving its positive side). The same is true for $k$-facets. This implies the following:

**Observation 6.1.** *The combinatorial structure of $k$-edges ($k$-facets, resp.) only changes during transitions of order $k$ and $k-1$.*

A crossing of $k$-edges can appear if some endpoint of a $k$-edge passes over another $k$-edge or if the combinatorial structure of the $k$-edges changes. From the discussion above, we can conclude:

**Observation 6.2.** *The number of crossings between $k$-edges only changes during transitions of order $k$ and $k-1$.*

The last unknown are the values $g_k$. Here, we are considering the usual $k$-facets in the underlying space $\mathbb{R}^3$. The only events in which $g_k$ could change are either the origin passing through a $k$-facet (by the earlier mentioned interpretation of the $g$-vector) or during a structural change of the $k$-facets.

The former case is exactly a transition of order $k$ (in the spherical sense). By Observation 6.1 we only need to worry about transitions (in the $\mathbb{R}^3$ sense) of order $k$ and $k-1$. A closer look at these (see Figure 6.2 on the next page) reveals that there is no topological change in the $k$-facet surface and hence, no change in the value $g_k$. This leads to:

**Observation 6.3.** *The value $g_k$ only changes during transitions of order $k$.*

Therefore, we can conclude that the only events when the quantities of interest do change are those of order $k$ and $k-1$.

### 6.1.3 Transition types

We have already classified transitions according to their order and type (convex and non-convex). In the previous subsection, we saw that only transitions of order $k$ and $k-1$ are relevant for us and we will look at them more closely in this subsection.

When talking about the transitions in the context of $k$-triangles, we will call them according to the shape formed by the participating $k$-triangles. We depict all the possible transitions in Figure 6.2 on the facing page; the $k$-triangles in the figure are cooriented upwards.

Let us start with convex transitions. In a convex transition of order $k$ (see Figures 6.2(a) to 6.2(b) on the next page), there are two $k$-triangles before and after the transition on the participating points and in both cases they form a roof-like shape (the $k$-triangles are attached to the lower "diagonal"). In a convex transition of order $k-1$ (see Figures 6.2(c) to 6.2(d) on the facing page), there are two $k$-triangles before and after the transition and they form a valley-like shape (the $k$-triangles are attached to the upper "diagonal"). Hence, we call the convex transitions a *roof transition* and a *valley transition*, respectively.

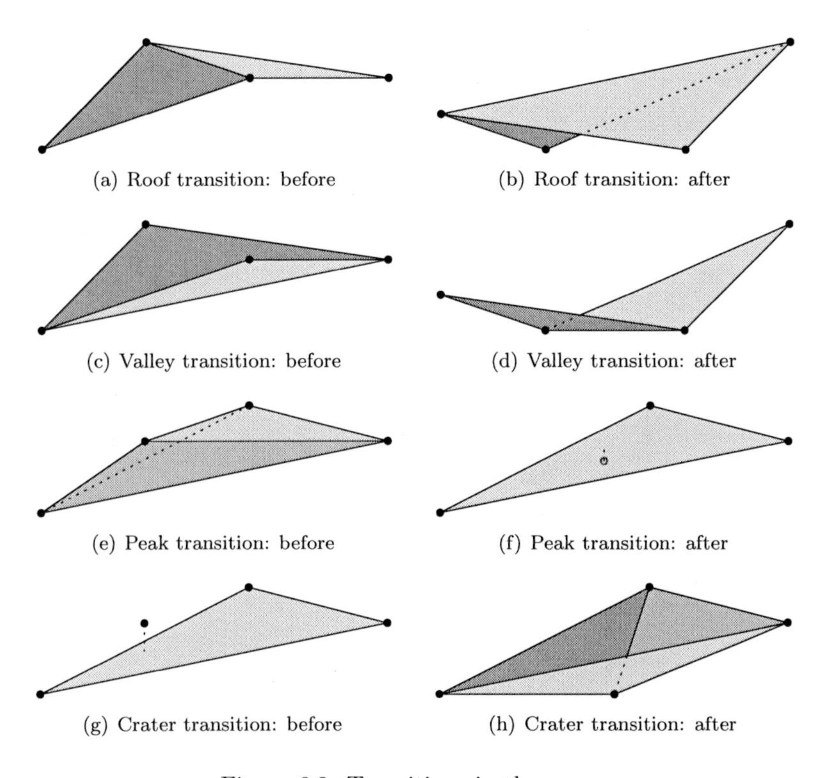

(a) Roof transition: before (b) Roof transition: after

(c) Valley transition: before (d) Valley transition: after

(e) Peak transition: before (f) Peak transition: after

(g) Crater transition: before (h) Crater transition: after

Figure 6.2: Transitions in the space

In a non-convex transition of order $k$ (see Figures 6.2(e) to 6.2(f) on the current page), we have three $k$-triangles from the "outer triangle" climbing up to the middle vertex (forming a peak) before the transition. This is replaced by the outer triangle as the middle point descends below its surface. In a non-convex transition of order $k-1$ (see Figures 6.2(g) to 6.2(h) on this page), there is only the outer triangle before but it gets replaced by the three $k$-triangles from its sides diving to the middle point (forming a crater) — as soon as the middle point sinks below the outer triangle. Hence, we call these transitions a *peak transition* and a *crater transition*, respectively.

We can adopt the same terminology for vector configurations on $\mathbb{S}^2$ (see Figure 6.3 on the next page). Recall that the origin $\mathbf{0}$ plays a special role here: only the $k$-facets in which it participates define spherical $k$-edges. Observe that during a non-convex transition, the point in the middle is always $\mathbf{0}$, so

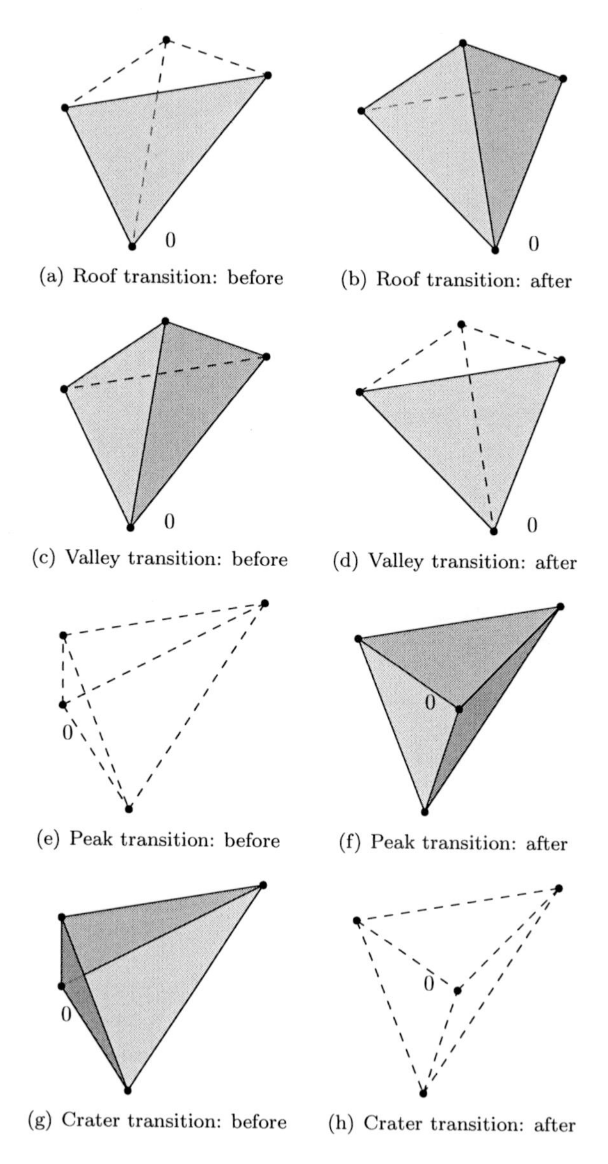


(a) Roof transition: before

(b) Roof transition: after

(c) Valley transition: before

(d) Valley transition: after

(e) Peak transition: before

(f) Peak transition: after

(g) Crater transition: before

(h) Crater transition: after

Figure 6.3: Convex transitions on the sphere

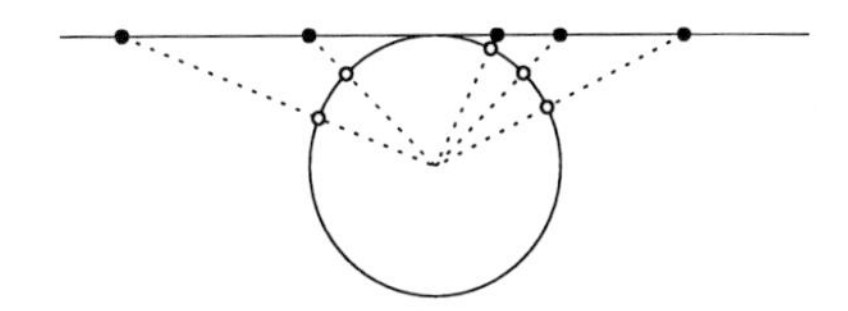

Figure 6.4: An example of the projection $q$ (in one dimension)

we can say that $\mathbf{0}$ *moves through a triangle* and for a convex transition that $\mathbf{0}$ *misses a triangle* (although $\mathbf{0}$ is fixed it is sometimes easier to imagine it as moving while the other points are fixed).

**Convention.** Each of the transitions has two directions: the one depicted in the figures and the reverse direction (i.e. going from the situation labelled "after" to the situation labelled "before"). It is always sufficient to study only one of the directions and we will stick to the one depicted in the figures.

### 6.1.4 On a hemisphere

Earlier in the text, we claimed that a point set on one hemisphere is equivalent to a planar point set. We make the statement more precise in the following definition and lemma.

Denote $\mathbb{S}^2_+ := \{x \in \mathbb{S}^2 \mid x_3 > 0\}$ the (open) *upper hemisphere* of the two-dimensional sphere and consider the plane $h := \{x \in \mathbb{R}^3 \mid x_3 = 1\}$. There is a natural bijection $q : \mathbb{S}^2_+ \to h$ defined by the stereographic projection $q(x) := \frac{x}{x_3}$, with the inverse mapping $q^{-1}(x) = \frac{x}{\|x\|_2}$.

In another words, the above projection is defined by shooting a ray from the origin to the upper half-space. Every such ray intersects the $\mathbb{S}^2_+$ as well as $h$ exactly once and these two points are identified by $q$ (see Figure 6.4).

Now, we have a notion of a particular hemisphere and a plane in $\mathbb{R}^3$ and can continue with the precise statement.

**Lemma 6.4.** *Let $V \subseteq \mathbb{S}^2_+$ be a set of $n$ vectors on the upper hemisphere and $P := \{q(v) \mid v \in V\}$ its projection onto the plane $h$. Then the following hold:*

*(i) a pair $(u, v) \in V^2$ forms a spherical $k$-edge on $\mathbb{S}^2$ if and only if the projections $(q(u), q(v)) \in P^2$ form a $k$-edge on the plane $h$,*

*(ii) two spherical $k$-edges $(s, t), (u, v) \in V^2$ cross if and only if the $k$-edges $(q(s), q(t))$ and $(q(u), q(v))$ cross.*

*Proof.* For the first part, it is sufficient to observe that every great circle $\ell$ of $\mathbb{S}^2$ restricted to the upper hemisphere $\ell_+ := \ell \cap \mathbb{S}^2_+$ gets projected onto a line (and vice versa). This is true because $\ell$ is an intersection of $\mathbb{S}^2_+$ with some linear hyperplane in $\mathbb{R}^3$ and therefore $q(\ell_+)$ is the intersection of $h$ with this hyperplane. Hence, for every pair of vertices the set of points in their positive hemisphere (in the setting on $V$) and in their positive half-plane (in the projection $P$) is the same modulo the bijection $q$. Similarly, hemispheres correspond to half-planes.

For the second part, recall that the spherical $k$-edges are drawn as the shorter arc. In our case this is always the one completely contained in the upper hemisphere and from the previous paragraph, we know that its image forms a line segment. Hence, the projections of spherical $k$-edges coincide with the $k$-edges in $h$ and vice versa. Since $q$ is a bijection between $\mathbb{S}^2_+$ and $h$, the crossings have to coincide, too. □

### 6.1.5 What does change?

This subsection discusses which of the values of interest ($e_k, g_k$ and the number of pinched crossings) undergo changes and when.

**$g$-vector.** We have already observed earlier that the values $g_k$ only change during a transition of order $k$ when the origin $\mathbf{0}$ passes through a $k$-triangle. This means in the above terminology that the change happens only during convex transitions – and as $\mathbf{0}$ leaves one layer of the $k$-facet surface, $g_k$ changes by $-1$.

**$e$-vector.** A brief look at the possible transitions (see Figure 6.3 on page 80) reveals that the number of $k$-edges changes by $+1$ and $-1$, respectively, during a convex transitions of order $k$ and $k-1$. During non-convex transitions of order $k$ and $k-1$, the changes are $+3$ and $-3$, respectively.

We can summarise these in the following:

**Observation 6.5.** *The change of the values $g_k, e_k$ and $e_{<k}$ during the transitions is characterised in the following table.*

| | *non-convex* | | *convex* | |
|---|---|---|---|---|
| | $k$ | $k-1$ | $k$ | $k-1$ |
| $g_{k-1}$ | 0 | $-1$ | 0 | 0 |
| $g_k$ | $-1$ | 0 | 0 | 0 |
| $e_k$ | $+3$ | $-3$ | $+1$ | $-1$ |
| $e_{<k}$ | 0 | $+3$ | 0 | $+1$ |

The remaining quantities of our interest will be investigated later.

## 6.2 Crossing identities

In this section, we present our results and theorems. Our ultimate goal is to study point sets in $\mathbb{R}^3$ and find an identity of similar nature to (5.1) which might (however, currently does not) help improving the upper bound on the number $e_k(P)$ in $\mathbb{R}^3$. The initial step in this direction is studying vector configurations on $\mathbb{S}^2$ which are the first generalisation of planar point sets.

We start by generalising the Lovász lemma to the sphere, which we prove in Theorem 6.6. Using that, we prove an identity (6.2) analogous to (5.1) in Theorem 6.7 on page 85. Both proofs are heavily based on continuous motion and it might be helpful to consult Figure 6.2 on page 79 and Figure 6.3 on page 80 a few times.

### 6.2.1 Spherical Lovász' Lemma

Let $V$ be a set of $n$ vectors on $\mathbb{S}^2$ in general position. The first thing to observe is that both Fact 5.2 and Fact 5.3 remain valid in this setting (instead of a line, one considers a great circle on the sphere)*. We will simply refer to these facts, even when using their $\mathbb{S}^2$ versions. We can prove a generalisation of the Lovász lemma.

**Theorem 6.6** (Lovász lemma on $\mathbb{S}^2$). *Let $V$ be a set of $n$ vectors on $\mathbb{S}^2$ in general position. Let $\ell$ be an oriented great circle on $\mathbb{S}^2$ avoiding the vectors of $V$. Denote by $\ell^+$ and $\ell^-$ the hemisphere on the right of $\ell$ and the hemisphere on the left of $\ell$, respectively. Then the number of (linear) $k$-edges intersected by $\ell$ is*

$$\begin{aligned} e_k(V,\ell) :=2 \cdot \big( &\min\{k+1, |V\cap \ell^+|, |V \cap \ell^-|\} \\ &+ g_{k-1}(V) - g_k(V)\big), \text{ for any } k < \frac{n-2}{2}. \end{aligned} \tag{6.1}$$

Note that this lemma relates linear $k$-edges (in the term $e_k(V,\ell)$) with $k$-facets of the underlying three-dimensional point set (the values $g_k(V)$ and $g_{k-1}(V)$).

*Proof.* To prove Theorem 6.6 we will proceed in three steps. First, we exhibit some vector configuration $V$ and a great circle $\ell$, for which (6.1) holds. Next, we show that if (6.1) holds for a configuration $V'$ with some great circle $\ell$, it

*By the interpretation of $k$-edges $pq$ on $\mathbb{S}^2$ as $k$-facets $pq\mathbf{0}$ in $\mathbb{R}^3$ we can simply project the point set onto the orthogonal complement of the line $\mathbf{0}p$ which is $\mathbb{R}^2$. Applying the facts to the image of $\mathbf{0}p$ under this projection yields their equivalents in $\mathbb{S}^2$.

also holds for $V'$ with every great circle (avoiding $V'$). Last, we show that moving points of $V$ preserves (6.1) as well which will conclude the proof.

If $V$ lies entirely in one open hemisphere of $\ell$, then it is easy to see that no $k$-edge intersects $\ell$ and (6.1) holds trivially.

Next, assume we have a fixed point set $V$ and some great circle $\ell$. We will show that moving $\ell$ continuously (while keeping $V$ fixed) can only change both sides of (6.1) by the same amount. Consider a continuous motion of $\ell =: \ell(0)$ to a final great circle $\ell(1)$. Assume the motion is generic, i.e. $\ell(t)$ always contains at most one vector, there are only finitely many instants when it does contain a vector and at each such instant, the vector in question changes the side of $\ell$ (locally, we can talk of the vector as moving).

The $g$-values are independent of $\ell$ and thus, do not change. The minimum as well as $e_k(P, \ell)$ is constant as long as the set $V \cap \ell^+$ remains unchanged. So the only changes happen when some point changes side of $\ell$ at an instant $t$. Then, the value $e_k(V, \ell(t))$ changes by $+2$ ($-2$, $0$, respectively) due to Fact 5.3 and the minimum then changes by $+1$ ($-1$, $0$, respectively) so the changes on both sides of (6.1) are equal.

For the last part, assume that we have a vector configuration $V$. We show that transforming $V =: V(0)$ to any other $V(1)$ in general position changes both sides of (6.1) by the same amount, by a standard continuous motion argument. The previous paragraph allows us to place the great circle $\ell$ anywhere we want throughout the motion.

As we already argued about points passing over the great circle (or vice versa), it is sufficient to study only the four possible transitions of order $k$ and $k-1$ and observe how $g_k$, $g_{k-1}$ and $e_k(V, \ell)$ change:

| | non-convex | | convex | |
|---|---|---|---|---|
| | $k$ | $k-1$ | $k$ | $k-1$ |
| $e_k(V, \ell)$ | +2 | -2 | 0 | 0 |
| $g_k(V)$ | -1 | 0 | 0 | 0 |
| $g_{k-1}(V)$ | 0 | -1 | 0 | 0 |

The claimed change in $g$-values is implied by Observation 6.5. It remains to examine $e_k(V, \ell)$. During any transition, only the edges of the triple that forms the critical great circle can change from being an $i$-edge to a $j$-edge for $i \neq j$. During a convex transition, this whole triple as well as the edges spanned by it lie on one open hemisphere. If $\ell$ is the great circle bounding this hemisphere, no crossings with $k$-edges can appear or disappear. During a non-convex transition, the three edges spanned by the coplanar triple form a great circle so they must intersect every other great circle $\ell$ in two points. It remains to see what happens with these edges. If the transition has order $m$, they are all $m$-edges before and become $(m+1)$-edges after the transition. All in all, we lose two crossings if $m = k$ and gain two if $m = k-1$.

If we compare the values in the table with the equation (6.1), we see that both sides always change by the same amount which concludes the whole proof. □

### 6.2.2 Pinched crossings

As we already mentioned earlier, every set $P$ of $n$ points in the plane satisfies the identity (5.1):

$$\mathrm{cr}_k(P) + \sum_{q \in P} \binom{\deg_k(q)}{2} = e_{<k}(P),$$

for every $0 \leq k < \frac{n-2}{2}$. For a point set $P \subset \mathbb{R}^3$ in convex position, the contraction $P/p$ (recall the definition on page 71) lies on one hemisphere and therefore, is equivalent to a planar point set (see Lemma 6.4). Thus, summing up the identities over all $p \in P$ (observe that $\sum_{p \in U} e_{<k}(U/p) = 3e_{<k}(U)$) yields

$$\mathrm{pcr}_k(P) + 2 \sum_{pq \in \binom{P}{2}} \binom{\deg_k(pq)}{2} = 3e_{<k}(P),$$

for all $0 \leq k < \frac{n-3}{2}$. We prove similar identities for edge crossings in a vector configuration on $\mathbb{S}^2$ and pinched crossings in a vector configuration on $\mathbb{S}^3$.

**Theorem 6.7.** *Let $V$ be configuration of $n$ vectors on the sphere $\mathbb{S}^2$, and let $0 \leq k < \frac{n-2}{2}$. Then*

$$\mathrm{cr}_k(V) + \sum_{q \in V} \binom{\deg_k(V/q)}{2} = e_{<k}(V) + m_k(V) \tag{6.2}$$

*where*

$$m_k(V) := (g_{k-1} - g_k)(2k + 1 + g_{k-1} - g_k) + 2g_{k-1}$$

*and $\mathrm{cr}_k, e_{<k}$ and $g_j$ denote $\mathrm{cr}_k(V), e_{<k}(V)$ and $g_j(V)$, respectively.*

Summing these up for contractions $U/p$ yields:

**Corollary 6.8.** *Let $U$ be a configuration of $n$ vectors on the sphere $\mathbb{S}^3$, and let $0 \leq k < \frac{n-3}{2}$. Then*

$$\mathrm{pcr}_k(U) + 2 \sum_{pq \in \binom{U}{2}} \binom{\deg_k(U/\{p,q\})}{2} = 3e_{<k}(U) + \sum_{p \in U} m_k(U/p) \tag{6.3}$$

*Proof of Theorem 6.7.* This proof is again a continuous motion argument.

If all the vectors of $V$ lie on one open hemisphere, then all $g_i$ are 0 and Lemma 6.4 together with the identity (5.1) implies the result. Every vector configuration can be obtained by a continuous deformation of a vectors lying on one hemisphere, so it is sufficient to analyse changes of the involved quantities under continuous motion.

We have already described when the changes in the numbers $g_i$ and $e_{<k}$ happen in Observation 6.5. The number of generalised crossings $\mathrm{cr}_k(V) + \sum_{q \in P} \binom{\deg_k(V/q)}{2}$ only changes when there is a change in the $k$-edges, i.e. during convex and non-convex transitions of order $k$ and $k-1$.

**Lemma 6.9.** *The following changes occur during the transitions mentioned below under the assumption $2k < n-2$:*

*(i) During a peak transition, the number $\mathrm{cr}_k + \sum_p \binom{\deg_k(p)}{2}$ of generalised crossings increases by $2(k + g_{k-1}^{\mathrm{old}} - g_k^{\mathrm{old}}) + 2$.*

*(ii) During a crater transition the number of generalised crossings decreases by $2(k + g_{k-1}^{\mathrm{old}} - g_k^{\mathrm{old}}) - 1$.*

*(iii) During a roof transitions, the number of generalised crossings does not change.*

*(iv) During a valley transition, the number of generalised crossings increases by 1.*

*Here, the upper index* $^{old}$ *refers to the quantity* before *the transition.*

*Proof.* We start with peak and crater transitions. Illustrations are given in Figure 6.5 on the facing page and Figure 6.6 on page 88. In both figures, we depict a view onto the sphere from above: the edges and vertices on the upper hemisphere are drawn by solid lines and filled circles, the edges on the lower hemisphere are drawn dashed and the points on the lower hemisphere are drawn as empty circles. We suggestively draw several possible $k$-edges not directly involved in the transition but contributing to some crossings which appear or disappear. We diverge from our usual notation and use an arrow to indicate the $k$-edges passing between the upper and the lower hemisphere (instead of indicating orientation). Such an edge is then drawn as two almost parallel segments with an arrow: one on the upper hemisphere (solid) and one on the lower hemisphere (dashed).

During a peak transition, the origin moves from the negative side of the affine $k$-facet $pqr$ of $V$ to its positive side. At the critical instant when $\mathbf{0}, p, q, r$ become coplanar, spherical $k$-edges $pq$, $qr$ and $rp$ forming a great circle are created. Thus, we gain exactly one generalised $k$-edge crossing for every branch

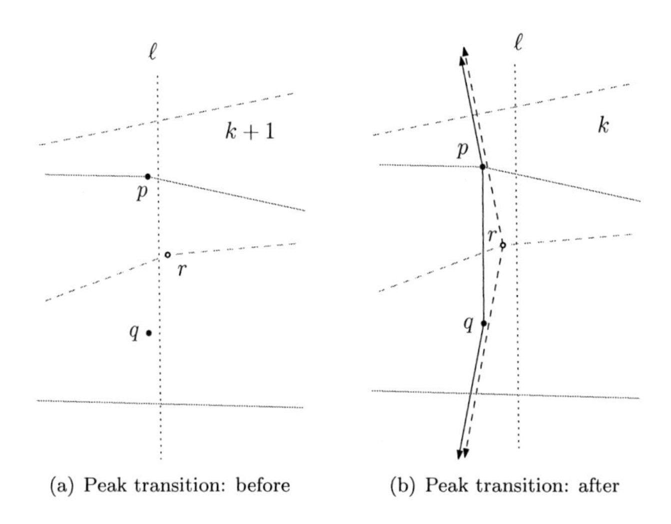


(a) Peak transition: before (b) Peak transition: after

Figure 6.5: Peak transition with a critical great circle.

of the $k$-edge curve which crosses the great circle $\ell$ (the critical great circle, appropriately tilted; see Figure 6.5).

Before the transition, there are $k+1$ points on one side of $\ell$ and $n-k-1$ points on the other side. Moreover, the assumption on $k$ ensures that $\min\{k+1, k+1, n-k-1\} = k+1$. Hence, by Theorem 6.6, the number of new crossings equals $2(k+1+g_{k-1}^{\mathrm{old}} - g_k^{\mathrm{old}})$. This proves the first part of the lemma.

In the crater transition, the situation is similar. The great circle $\ell$ depicted in Figure 6.6 on the next page intersects exactly $2(k + g_{k-1}^{\mathrm{old}} - g_k^{\mathrm{old}})$ $k$-edges. There we count four edges, which are not crossed by $pq$, $qr$ or $rp$ but on the other hand we do not count three degenerate crossings of $pqr$ in the vertices $p, q$ and $r$. Thus, there are exactly $2(k + g_{k-1}^{\mathrm{old}} - g_k^{\mathrm{old}}) - 1$ generalised crossings of $pqr$ which disappear after the transition. This proves the second part of the lemma.

The analysis of the convex transitions is identical to the analysis in the planar version of the identity (see Figure 6.7 on the following page). During a roof transition there is the same number of $k$-edges emanating to both sides of the critical great circle $\ell$ from each of the points $p, q, r$ involved in the transition. Consequently, there is no change in the total number of generalised crossings (only some proper crossings become degenerate).

During a valley transition, the middle vector loses one branch of the $k$-edge curve passing through it but on the other hand a new $k$-edge starts intersecting

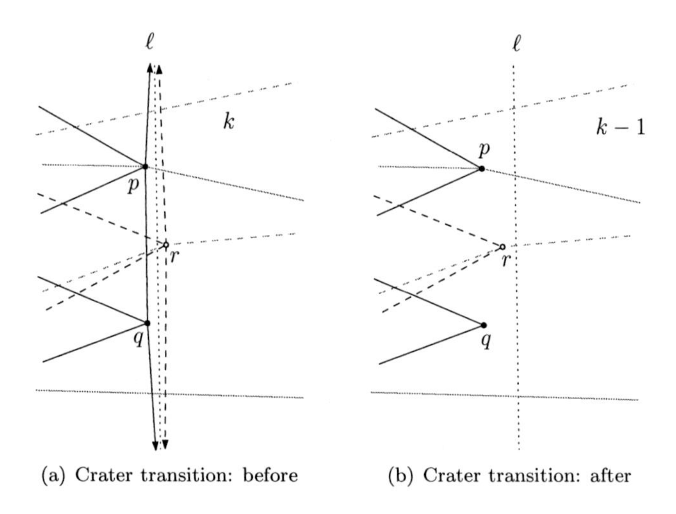


(a) Crater transition: before (b) Crater transition: after

Figure 6.6: Crater transition with a critical great circle.

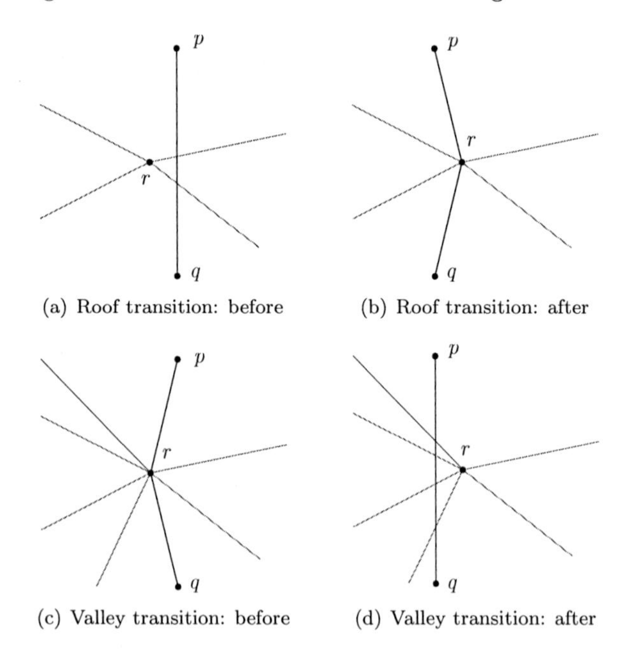


(a) Roof transition: before (b) Roof transition: after

(c) Valley transition: before (d) Valley transition: after

Figure 6.7: Changes in roof and valley transitions.

one more $k$-edge than there are branches through the middle vector. Therefore, the number of crossings increases by 1. □

Let us set $X := \mathrm{cr}_k(V) + \sum_{q \in P} \binom{\deg_k(V/q)}{2}$ and $\Delta g_k := g_{k-1} - g_k$. We summarise the changes discussed above in the upper part of Table 6.1. To find

| | non-convex | | convex | |
|---|---|---|---|---|
| | $k$ | $k-1$ | $k$ | $k-1$ |
| $g_{k-1}$ | $0$ | $-1$ | $0$ | $0$ |
| $g_k$ | $-1$ | $0$ | $0$ | $0$ |
| $e_{<k}$ | $0$ | $-3$ | $0$ | $+1$ |
| $X$ | $+2(k + \Delta g_k^{\mathrm{old}}) + 2$ | $-2(k + \Delta g_k^{\mathrm{old}}) + 1$ | $0$ | $+1$ |
| $X - e_{<k}$ | $+2(k + 1 + \Delta g_k^{\mathrm{old}})$ | $-2(k + 1 + \Delta g_k^{\mathrm{old}})$ | $0$ | $0$ |
| $\Delta g_k$ | $+1$ | $-1$ | $0$ | $0$ |
| $(\Delta g_k)^2$ | $+2\Delta g_k^{\mathrm{old}} + 1$ | $-2\Delta g_k^{\mathrm{old}} + 1$ | | |
| $(2k+1)\Delta g_k$ | $+2k + 1$ | $-2k - 1$ | | |
| $2g_{k-1}$ | $0$ | $-2$ | | |

Table 6.1: Changes in the various transition (upper part) and useful expressions (middle and lower parts).

the desired identity for Theorem 6.7, let us first note that from the convex transition of order $k-1$ we need to have an identity of the form $\mathrm{cr}_k(V) + \sum_{q \in P} \binom{\deg_k(V/q)}{2} - e_{<k} = f(g_k, g_{k-1})$. Furthermore, the $g$-values do not change during convex transitions, thus the choice of $f$ will not have any effect on the change in these transitions. Hence, the key is finding $f$ which compensates for the change in the non-convex transitions.

As the lower part of Table 6.1 suggests, we can simply "build" the function $f$ from the building blocks $(\Delta g_k)^2$, $\Delta g_k$ and $g_{k-1}$ as the combination $(\Delta g_k)^2 + (2k+1)\Delta g_k + 2g_{k-1}$. This is the same quantity as $m_k$ claimed by the theorem and therefore, concludes the proof. □

## 6.3 Concluding remarks

Using continuous motion arguments, we have derived a crossing identity for the number of pinched crossings between pairs of $k$-facets in $\mathbb{R}^3$. It would also be interesting to obtain an identity of a similar spirit for the number of *crossing triples* of $k$-facets.

Dey's bound $e_k \leq O(n^{4/3})$ for point sets in the plane follows by combining the upper bound $\mathrm{cr}_k = O(kn)$ with the crossing lemma of Ajtai et al. [ACNS82] and Leighton [Lei84]. Theorem 6.7 immediately gives an upper bound on the number $\mathrm{cr}_k \leq n(k+1)+m_k$ (as $e_k \leq n(k+1)$ by the analysis of Clarkson and Shor [CS89]).

**Corollary 6.10.** *Let $V$ be a set of $n$ vectors on the sphere $\mathbb{S}^2$. Then $e_k(V) \leq O(\sqrt[3]{n^4 + n^2 m_k(V)})$.*

For vector sets lying on one open hemisphere (i.e. equivalent to planar point set), this coincides with the Dey's $k$-edge upper bound $O(n^{4/3})$ and as $m_k$ grows, it gets to the trivial $O(n^2)$ upper bound for the maximal possible values of $m_k$. Notice, that one can not really hope for a better upper bound in general since there are point sets on the sphere with quadratically many halving edges (a point set obtained from a point set on the $d-1$ dimensional moment curve in the hyperplane $x_d = 1$ by projecting the points through the origin $\mathbf{0}$ alternatingly to the upper and lower hemisphere of the unit sphere $\mathbb{S}^{d-1}$ in $\mathbb{R}^d$).

There is still one burning question (not) to be answered: does our result imply anything new for the $k$-sets in $\mathbb{R}^3$? The answer is unfortunately: we do not know whether or how.

The hope of our approach would be an improved upper bound on the number of pinched crossings from Corollary 6.8. We miss, however, any means of controlling the sum of $m_k$. Currently, we cannot do any better than using an upper bound of Sharir et al. [SST00, SST01] on the number of crossings (which is actually what we wanted to bound).

# 7 Crossing bounds

We already mentioned one specific kind of higher dimensional crossing, the pinched crossings, earlier. In Chapter 7, we meditate for a while on the question of an appropriate definition of a higher dimensional crossing, namely on an analogue of the rectilinear crossing number of complete graphs. Out of two reasonable definitions of crossings — one being intersecting simplices of higher dimension and the other being convex $d+2$ point subsets — we study the weaker one (the latter one dealing with convex subsets) and we prove non-trivial upper and lower bounds on the minimum number of crossings in higher dimensional point sets.

In Section 7.1, we consider the definitions of crossings in the plane and discuss several of the their possible natural generalisations. Only two of those do make sense and we establish the generalisation of the rectilinear crossing number to higher dimensions with respect to these definitions of crossings.

In Section 7.2, we prove lower bounds on these analogues of the rectilinear crossing number. We start by a very straightforward averaging argument which gives a lower bound for both notions of crossings and later, we use the ideas from Lovász et al. [LVWW04] to improve this bound for the convex-subset notion of crossings.

In Section 7.3, we give an upper bound on the number of generalised crossings but the lower and upper bounds are still far apart.

This is unpublished work.

## 7.1 What are crossings?

The *rectilinear crossing number* of a graph $G$ is the minimum number of crossings in a straight-edge drawing of $G$ in the plane (i.e. a drawing, where all edges are straight-line segments). A case which is especially interesting is the complete graph $K_n$. One defines $\overline{\text{cr}}(P)$ of a point set $P$ in general position to be the number of crossings in a straight line drawing of the complete graph on $P$, i.e. the number of (proper) crossings of all the segments connecting two points of $P$. Since there is a crossing pair of segments on given four points if and only if they lie in convex position, $\overline{\text{cr}}(P)$ is also the number of convex quadrilaterals in $P$. Typically, one is interested in the quantity $\overline{\text{cr}}(n) := \min_{P \subseteq \mathbb{R}^2, |P|=n} \overline{\text{cr}}(P)$ where again only point sets $P$ in general position are considered.

What is the right generalisation of a crossing in higher dimensions? How can we generalise the crossing number $\overline{\text{cr}}(P)$ of a point set?

For this to make sense, we should better seek for a $d$-dimensional analogue $\overline{\text{cr}}_d(P)$ which attains non-zero value for all sufficiently large point sets. Then there are several possibilities:

**Intersecting simplices.** In dimensions 3 and higher, there is no hope we can make two segments intersect: if the point set $P$ is in general position, this will never happen. One possibility is to consider higher-dimensional simplices instead. On the one hand, a simple dimensionality argument implies that in $\mathbb{R}^d$, one needs $k$-dimensional simplices (i.e. spanned by $k+1$ points) with $2k \geq d$ if there should be any intersections (as otherwise, the affine hulls do not intersect if the point set is in general position).

On the other hand, the Van Kampen-Flores theorem [vK32, Flo33] (in a weaker geometric version) says the following:

**Theorem 7.1** (Geometric Van Kampen-Flores). *For any $k \geq 1$ and any set $P$ of $2k+3$ points in $\mathbb{R}^{2k}$ there are two disjoint subsets $A, B \subseteq P$ of $|A| = |B| = k+1$ whose convex hulls have a non-empty intersection.*

This implies that a set of $d+3$ points in $\mathbb{R}^d$ contains two disjoint subsets $A$ and $B$ with $|A| = \lceil \frac{d}{2} \rceil + 1$ and $|B| = \lfloor \frac{d}{2} \rfloor + 1$, whose convex hulls intersect.

In even dimensions, this is exactly the Van Kampen-Flores theorem. For odd dimensions one needs to consider the Van Kampen-Flores theorem for $\mathbb{R}^d$ embedded as the hyperplane $x_{d+1} = 0$ in $\mathbb{R}^{d+1}$ and an additional artificial point with a non-zero last coordinate. This point is redundant for any intersection purposes. The Van Kampen-Flores theorem for $\mathbb{R}^{d+1}$ guarantees two intersecting simplices of cardinality $\lceil \frac{d}{2} \rceil + 1$. If one of them contains the artificial point, this can be removed since the other simplex completely lies in $x_{d+1} = 0$ and so does the intersection. Otherwise, we have two disjoint subsets $A, B$ of $\lceil \frac{d}{2} \rceil + 1$ points in $\mathbb{R}^d$ that intersect but we still need to remove one of the points. Recall, that the points are in general position (reminder: we are now back in $\mathbb{R}^d$) and hence, the intersection of their affine hulls $\mathrm{aff}(A) \cap \mathrm{aff}(B)$ has dimension $d - 2\lfloor \frac{d}{2} \rfloor = 1$ and consequently, is a line $\ell$. Since $\ell \subseteq \mathrm{aff}(A)$ and $\ell \subseteq \mathrm{aff}(B)$, the line $\ell$ also intersects some proper faces of $\mathrm{conv}(A)$ and $\mathrm{conv}(B)$. Thus, the intersection $\mathrm{conv}(A) \cap \mathrm{conv}(B)$ is a (possibly degenerate) line segment each of whose endpoints is an intersection of $\ell$ with a proper face of one of the simplices $\mathrm{conv}(A), \mathrm{conv}(B)$. Taking the other simplex and the face yields two intersecting simplices of desired sizes.

This suggests the following definition:

**Definition 7.2** (Simplicial crossing). *For a set $P$ of points in $\mathbb{R}^d$ in general position, define $\overline{\mathrm{cr}}_d(P)$ to be the number of pairs of disjoint subsets $A, B \subseteq P$ with $|A| = \lceil \frac{d}{2} \rceil + 1$ and $|B| = \lfloor \frac{d}{2} \rfloor + 1$ whose convex hulls intersect. We call such a pair of sets $A, B$ a* simplicial crossing. *We again define $\overline{\mathrm{cr}}_d(n) := \min_{P \subseteq \mathbb{R}^d, |P|=n} \overline{\mathrm{cr}}_d(P)$.*

One might also insist on equally sized simplices in odd dimensions as well: then the solution is asking for two simplices on $\lceil \frac{d}{2} \rceil + 1$ vertices.

The Van Kampen-Flores theorem implies that $\overline{\mathrm{cr}}_d(d+3) \geq 1$. An analogous notion of crossings can be defined for arbitrary $(k+1)$-uniform hypergraphs embedded in $\mathbb{R}^{2k}$. In $\mathbb{R}^3$ we ask for an intersection of a segment and a triangle spanned by the point set: see Figure 7.1 on page 95 for illustration.

**Convex subsets.** Another possible viewpoint is the convex quadrilateral interpretation. In $d = 2$ we were asking for the convex $(2+2)$ tuples of points. We can do something similar in higher dimensions.

**Definition 7.3** (Convex crossings). *For a set $P$ of points in $\mathbb{R}^d$ in general position, define $\bigcirc_d(P)$ to be the number of $(d+2)$-point subsets of $P$ in convex position. Denote $\bigcirc_d(n) := \min_{P \subseteq \mathbb{R}^d, |P|=n} \bigcirc_d(P)$.*

This is the generalisation we will mainly be interested in. A set of $d+2$ points in $\mathbb{R}^d$ is either in convex position or forms a simplex with a point inside. The latter has only one partition into two subsets whose convex hulls intersect: into the simplex and the point inside. This also implies that two subsets of cardinalities $\lceil \frac{d}{2} \rceil + 1$ and $\lfloor \frac{d}{2} \rfloor + 1$ with intersecting convex hulls form together a set of $d+2$ points in convex position. Hence, the Van Kampen-Flores theorem also implies that any $d+3$ points contain a subset of $d+2$ points in convex position.* Moreover, the Radon partition is unique if the points are in general position and hence, every $d+2$ points in convex position contain at most one simplicial crossing. Consequently, we have $\bigcirc_d(P) \geq \overline{\mathrm{cr}}_d(P)$ and $\bigcirc_d(n) \geq \overline{\mathrm{cr}}_d(n)$.

**The other options.** In the previous paragraph, we considered two possibilities: either requiring two almost equally sized simplices that intersect (which is a particular type of a set of points in convex position) or just any subset in convex position. There are several other "types" of $d+2$ point sets in convex position according to their Radon partition: we know that every set of $d+2$ points in $\mathbb{R}^d$ has a Radon partition and it is unique for point sets in general position which allows one to classify the point sets according to the sizes (dimensions) of the simplices in the Radon partition. The question is, whether asking for some "other" specific convex position would make sense as well. The answer is: no. In a set of points on the moment curve, any $\lfloor \frac{d}{2} \rfloor$ points form a facet of the convex hull and hence, are disjoint from a convex hull of any other subset of points on the moment curve. Consequently, the smaller simplex has to contain at least $\lfloor \frac{d}{2} \rfloor + 1$ points in order to get a meaningful definition.

**Crossings in low dimensions.** Observe that both the notions of crossings coincide in $\mathbb{R}^2$ and $\mathbb{R}^3$. In the plane, it is by design but we would like to say a few words about $\mathbb{R}^3$. It is not too hard to see as a consequence of Radon's lemma that five points in convex position in $\mathbb{R}^3$ form a triangle intersected by a segment (see Figure 7.1 on the facing page). As one can see from the picture, these crossings are closely related to the pinched crossings from earlier (a segment-triangle crossing defines three pinched crossings and vice versa a pinched crossing induces a segment intersecting a triangle).

## 7.2 Lower bounds

As we indicated earlier, we will mainly examine the number $\bigcirc_d(n)$. Before we get to do that, we start with a very easy lower bound which works for $\overline{\mathrm{cr}}_d(n)$ as well as $\bigcirc_d(n)$.

---

* As long as $d > 1$. This fact is also easy to prove without such heavy machinery as Van Kampen-Flores theorem. We will see this in Observation 7.4.

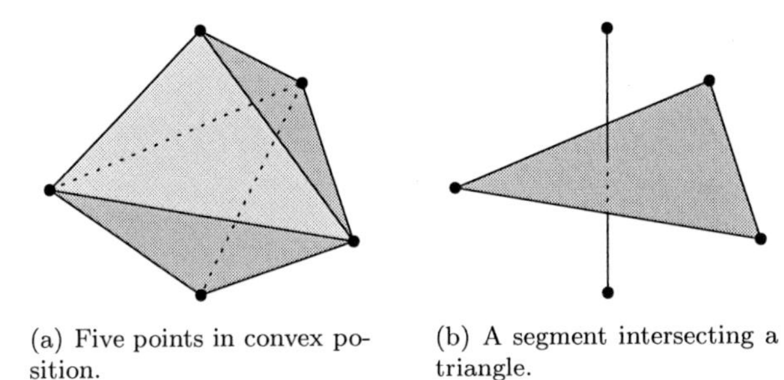

(a) Five points in convex position.

(b) A segment intersecting a triangle.

Figure 7.1: Points in convex position versus a segment intersecting a triangle.

### 7.2.1 Easy bound

We have already noticed that any $d+3$ points contain a simplicial crossing and therefore also a subset of $d+2$ points in convex position. For completeness we start with an elementary proof of the latter consequence.

**Observation 7.4.** *Let $P$ be a set of $d+3$ points in $\mathbb{R}^d$ in general position. Then it contains a subset of $d+2$ points in convex position.*

*Proof.* Pick any convex hull vertex $p \in P$ and consider the set $Q := P \setminus \{p\}$. If $Q$ is in convex position, we are done. Otherwise there are $d+1$ points on the convex hull and a point $q \in \mathrm{int}(\mathrm{conv}(Q))$. But then the segment $pq$ intersects some facet $F$ of the convex hull $\mathrm{conv}(Q)$ and hence, the points $p$ and $q$ together with the vertices of $F$ are in convex position. □

So we know that any $d+3$ points induce a crossing (either in the sense of a simplicial crossing or of a convex subset). We will give a quick lower bound by an averaging argument.

Denote by $t$ the number of pairs of sets $A, B$, where $|A| = d+2, |B| = d+3$, such that $A \subseteq B$ and $A$ is in convex position. The above observations yields

$$t \geq \binom{n}{d+3},$$

and on the other hand, every set of $d+2$ points in convex position can be extended in $n-d-2$ possible ways, and consequently

$$(n-d-2) \cdot \bigcirc_d(P) = t \geq \binom{n}{d+3}.$$

This implies the bound

$$\bigcirc_d(n) \geq \frac{1}{d+3} \cdot \binom{n}{d+2}.$$

An analogous argument for the simplicial crossings (using the Van Kampen-Flores theorem) yields:

$$\overline{\mathrm{cr}}_d(n) \geq \frac{1}{d+3} \cdot \binom{n}{d+2}.$$

### 7.2.2 Convex subsets

To construct a better lower bound, we can seek inspiration in the paper of Lovász et al. [LVWW04], where the authors give a lower bound on the rectilinear crossing number.

We do a similar calculation in higher dimensions. Consider a set $P$ of $n$ points in $\mathbb{R}^d$ in general position. We already defined $\bigcirc_d(P)$ as the number of $d+2$ point subsets in convex position. Let us denote $\triangle_d(P)$ the number of $d+2$ point subsets in non-convex position, i.e. forming a simplex with a point inside.

On the one hand, we know that any subset of $d+2$ points is either in convex or in non-convex position and thus,

$$\bigcirc_d(P) + \triangle_d(P) = \binom{n}{d+2}. \tag{7.1}$$

On the other hand, we can count the number $t$ of triples $(h, x, y)$, where $h$ is a cooriented hyperplane defined by exactly $d$ points of $P$, $x \in P$ lies in the positive half-space $h^+$ and $y \in P$ in the negative half-spaces $h^-$.

We know that every $h$ is defined by a $k$-facet of $P$ for some value $k$ and it can be completed to such a triple in $k \cdot (n-k-d)$ ways which gives

$$t = \sum_{i=0}^{n-d} e_i \cdot i \cdot (n-i-d). \tag{7.2}$$

Moreover, each such triple is defined by $d+2$ points and it is sufficient to see how many such triples a set of $d+2$ points gives rise to. The number of triples is exactly the number of (oriented) halving facets of the point set. Suppose that $A_d$ and $B_d$ are lower bounds on the number of halving facets of a set of $d+2$ points in convex and non-convex position, respectively, and that $A_d < B_d$, which leads to

$$A_d \bigcirc_d (P) + B_d \triangle_d(P) \leq t. \tag{7.3}$$

We will return to the actual values of $A_d$ and $B_d$ later. For now, we express a lower bound on $\bigcirc_d(P)$ in terms of $A_d$ and $B_d$.

**Reformulating (7.2).** We will find it useful to have (7.2) in a different form. First, notice that the formula $(a+b)\cdot(a-b) = a^2 - b^2$ implies

$$(n-i-d)\cdot i = \left(\frac{n-d}{2}\right)^2 - \left(\frac{n-d}{2} - i\right)^2.$$

Substituting into (7.2) and observint that $\sum_{i=0}^{n-d} e_i = 2\binom{n}{d}$, we obtain:

$$\sum_{i=0}^{n-d} e_i \cdot i \cdot (n-i-d) = \underbrace{2\binom{n}{d}\left(\frac{n-d}{2}\right)^2}_{\binom{d+2}{2}\binom{n}{d+2}+O(n^{d+1})} - \underbrace{\sum_{i=0}^{n-d} e_i \cdot \left(\frac{n-d}{2} - i\right)^2}_{=:X_d \geq 0}. \tag{7.4}$$

**Lower bounding $\bigcirc_d(n)$.** Combining the equations (7.3) and (7.4) yields:

$$A_d \bigcirc_d (P) + B_d \triangle_d(P) \leq \binom{d+2}{2}\binom{n}{d+2} - X_d + O(n^{d+1}).$$

At this point, subtracting a $B_d$ multiple of (7.1) results in

$$(A_d - B_d) \bigcirc_d (P) \leq \left(\binom{d+2}{2} - B_d\right)\binom{n}{d+2} - X_d + O(n^{d+1}),$$

and since $A_d - B_d$ is negative, this implies

$$\bigcirc_d (P) \geq \frac{1}{B_d - A_d}\left(B_d - \binom{d+2}{2}\right)\binom{n}{d+2} + \frac{X_d}{B_d - A_d} + O(n^{d+1}). \tag{7.5}$$

We are only interested in the leading term of $\bigcirc_d(P)$. For the time being, let us ignore the term $X_d \geq 0$. Carefully examining it will yield an improvement, but this will become negligible in high dimensions.

**Calculating $A_d$ and $B_d$.** It is time to find the values $A_d$ and $B_d$. Let us start in $\mathbb{R}^3$ where the life is sufficiently simple. Recall again that $A_d$ and $B_d$ is (a lower bound on) the number of (oriented) halving facets of a $d+2$ point set in convex and non-convex position, respectively. For a set of points in convex position, $A_3 = 2 \cdot 4$ and in non-convex position, $B_3 = 2 \cdot 6$ (the middle point together with any edge of the simplex).

We can plug this into (7.5) and obtain the lower bound

$$\bigcirc_3(n) \geq \frac{1}{2}\binom{n}{5} + O(n^4).$$

This can be further improved by lower bounding $X_3$ but we will return to this later.

In higher dimensions, there is one more difficulty that we need to overcome: we cannot calculate the number of halving facets in convex position directly because there are several different types of convex position. However, observe the following: as set of $d$ points among $d+2$ points either (i) is a 0-facet or a 2-facet, i.e. a convex hull facet, or (ii) is a 1-facet (which is halving). By the upper bound theorem [McM70], the number of convex hull facets is at most

$$f_{d-1} \leq \binom{(d+2) - \lfloor \frac{d+1}{2} \rfloor}{(d+2) - d} + \binom{(d+2) - \lfloor \frac{d+2}{2} \rfloor}{(d+2) - d} = \binom{\lceil \frac{d+3}{2} \rceil}{2} + \binom{\lceil \frac{d+2}{2} \rceil}{2}.$$

Hence, the total number of halving facets is at least

$$A_d = 2\underbrace{\binom{d+2}{d}}_{=\binom{d+2}{2}} - 2\binom{\lceil \frac{d+3}{2} \rceil}{2} - 2\binom{\lceil \frac{d+2}{2} \rceil}{2}.$$

As for non-convex position, it is easy to see that halving facets are exactly those $d$-point subsets formed by the point in the middle and another $d-1$ points of the simplex. Hence,

$$B_d = 2\binom{d+1}{2}.$$

**Finishing the calculation.** Some time spent calculating yields the following:

$$B_d - \binom{d+2}{2} = \frac{d^2 - d - 2}{2} \quad \text{and} \quad B_d - A_d = \begin{cases} \frac{d^2-1}{2} & d \text{ odd}, \\ \frac{d^2}{2} & d \text{ even}. \end{cases}$$

This implies the final bound:

**Theorem 7.5.** *Let $n \in \mathbb{N}$. Then the following holds:*

$$\bigcirc_d(n) \geq \gamma_d \binom{n}{d+2} + O(n^{d+1}), \quad \textit{where} \quad \gamma_d = \begin{cases} 1 - \frac{1}{d-1} & d \textit{ odd}, \\ 1 - \frac{d+2}{d^2} & d \textit{ even}. \end{cases}$$

Let us just check that this is true for $d = 3, 4$. We already calculated $\gamma_3 = 1/2$ which matches the general bound. For $d = 4$ we have $A_4 = 2 \cdot (15 - 6 - 3) = 12$ and $B_4 = 20$. Thus, $\gamma_4 = (20 - 15)/(20 - 12) = 5/8$ which also matches the formula.

**Using $X_d$.** Until now, we ignored the term $X_d$, but a further improvement is possible by lower bounding this term. Substituting $e_k = e_{\leq k} - e_{\leq k-1}$ into the definition of $X_d$ in (7.4) yields

$$X_d = 2 \cdot \sum_{i=0}^{(n-d)/2} e_{\leq i} \cdot \underbrace{\left( \left( \frac{n-d}{2} - i \right)^2 - \left( \frac{n-d}{2} - i - 1 \right)^2 \right)}_{(n-d-2i-1)}$$

There is a lower bound on $e_{\leq k}$ in arbitrary dimension due to Aichholzer et al. [AGOR09] which states that

$$e_{\leq k} \geq (d+1) \binom{k+d}{d}, \tag{7.6}$$

for $k$ up to $\frac{n}{d+1}$. Beyond this point, one can only use the monotonicity of $e_{\leq k}$ or some very minor improvements in certain ranges.

Calculations yield an improved bound in $\mathbb{R}^3$:

$$\bigcirc_3(n) \geq \frac{9}{16} \binom{n}{5} + O(n^4) \approx 0.563 \binom{n}{5} + O(n^4).$$

Moreover, if we let the dimension grow (slower than $n$ grows), the improvement of the constant in front of $\binom{n}{d+2}$ becomes of the order $1/d^d$ and is therefore almost negligible. Even if the bound (7.6) were true in the whole range of $k$, we would still only get an improvement of roughly $1/2^d$ in $\mathbb{R}^d$ and $1/8$ instead of $1/16$ in $\mathbb{R}^3$.

### 7.2.3 Remarks

We gave a lower bound on the crossing number $\bigcirc_d(n)$ of close to $(1-1/d)\binom{n}{d+2}$. Although we proceeded similarly as Lovász et al. [LVWW04], unlike them we do not make much use of the lower bound on the numbers $e_{\leq k}$.

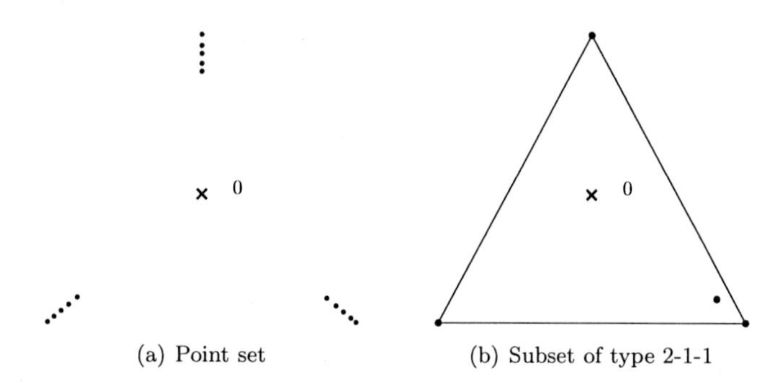


(a) Point set (b) Subset of type 2-1-1

Figure 7.2: A tripod construction in the plane

Notice that use of the upper bound theorem tends to be tight in the cases, when there are also many ($\leq k$)-facets. This suggests that there might be some balance between the two terms on the right hand side of (7.5) which one should try to exploit.

In our approach, we used two equations (or inequality in one case) involving two variables $\bigcirc_d(n)$ and $\triangle_d(n)$ to determine $\bigcirc_d(n)$. Since there are $\lfloor \frac{d}{2} \rfloor$ different types of convex position in $\mathbb{R}^d$, one might try to consider more equations of similar type (i.e. depending on the values $e_{\leq k}$) to estimate the sum of the numbers of different types of convex position. We are quite sceptical about this approach, though.

## 7.3 Upper bound

We will see that this lower bound is quite far from the natural upper bound which generalises the planar tripod construction (see Figure 7.2).

Consider a simplex $\sigma_d$ in $\mathbb{R}^d$, whose centre of gravity lies at the origin $\mathbf{0}$. Replace each of the $d+1$ vertices $v_i$ of the simplex by a small line segment connecting the point $(1-\varepsilon)v_i$ and $v_i$, i.e. a tiny segment directed towards the origin. Now put $k$ points $\{p_i^j\}_{j=1}^k$ on this segment. We will say that $p_i^j$ belongs to the vertex $v_i$ of the simplex. This yields a set $P = \{p_i^j\}_{(i,j)\in[d+1]\times[k]}$ of $n = k \cdot (d+1)$ points which is not in general position but an arbitrarily small perturbation $P'$ of $P$ is.

The $d+2$ point subsets can be classified according to the number of points belonging to each vertex. We observe that a subset $S$ of $d+2$ points which has exactly two points belonging to one vertex of the simplex and one point to each other vertex is in non-convex position. Out of the two points of $S$ belonging to the same vertex, denote the one closer to the origin by $x$ and the other one by $y$. Then $S - x$ forms a simplex. Assuming that the perturbation $P'$ of $P$ was sufficiently small, the centre of gravity of this simplex lies arbitrarily close to $\mathbf{0}$. Furthermore, the ray from $y$ through $x$ passes also arbitrarily close to $\mathbf{0}$ and hence, also arbitrarily close to the centre of the simplex. This implies that $x \in \operatorname{conv}\{S \setminus \{x\}\}$.

Let us calculate, how many such subsets there are:

$$\triangle_d(P') \geq \frac{d+1}{2} \cdot \left(\frac{n}{d+1}\right)^{d+2} + O(n^{d+1}) = \frac{(d+2)!}{2(d+1)^{d+1}} \binom{n}{d+2} + O(n^{d+1}).$$

This is of the order $e^{-d}\mathrm{poly}(d)$. It immediately implies an upper bound on $\bigcirc_d(n)$. For $d = 3$ we get:

$$\bigcirc_3(n) \leq \frac{49}{64} \binom{n}{5} + O(n^4).$$

So even in $\mathbb{R}^3$, the gap is still quite large since the lower bound constant is $\frac{36}{64}$ and it would still be large even if the $\leq k$-facet lower bound (7.6) was valid everywhere (then the lower bound would give $\frac{40}{64}$).

# Part III

# Conflict-free colouring

# 8 Conflict-free colourings: background

The whole Part III deals with yet another geometric topic, conflict-free and unique-maximum colourings of (mostly geometric) hypergraphs.

A special case of conflict-free colouring, called ordered colouring of graphs, has already been studied in 1988 by Iyer et al. [IRV88] and in 1995 by Katchalski [KMS95], but the notion of conflict-free colouring in its full generality entered the scene only very recently, in 2003, in the works of Even et al. [ELRS03] and Smorodinsky [Smo03]. These two works led to a burst of research on all possible conflict-free fronts: starting from algorithms for conflict-free colouring geometric and other hypergraphs with only few colours, through online variants of the problem, to some very special settings such as shallow regions or generalisations and specialisations of conflict-free colourings. It goes without saying that many famous notions and theorems, such as VC-dimension or the algorithmic local lemma, appear in several of the above mentioned results and their proofs.

Our goal for this chapter is to give an overview of some of the most important notions and results in this area. We include the observations and lemmata we find useful and will need in later chapters, as well as the results which may help the reader place our work in the context of the current research.

In the subsequent chapters, we present our work on list conflict-free colourings and list unique-maximum colourings which are a slightly more restricted variant thereof. In Chapter 9 we design a general algorithm for list unique-maximum colouring of arbitrary hypergraphs using a new potential method. As corollaries, we obtain asymptotically optimal bounds for list colouring many hypergraphs arising in geometric setting. In Chapter 10 we include several smaller results for list conflict-free colouring: bounds on some online variants of the list conflict-free colouring problems, asymptotically optimal bounds on the list conflict-free colouring of graphs with respect to paths, as well as a general relationship of the conflict-free choice number to the conflict-free chromatic number.

## 8.1 Definitions and basics

Before we dive into the topic of conflict-free colouring, let us start with the basic definitions and notations that will be used throughout the following chapters.

### 8.1.1 Hypergraphs

We start with basic definitions of a hypergraph, independent set in a hypergraph and an induced subhypergraph.

**Definition 8.1** (Hypergraph). *A hypergraph is a pair $H = (V, \mathcal{E})$, where $\mathcal{E} \subseteq 2^V$. Elements of $V$ are called* vertices *of $H$ and elements of $\mathcal{E}$ are called* hyperedges *of $H$. We write $V(H)$ for the vertices of $H$ and $\mathcal{E}(H)$ for the hyperedges of $H$ if they are not explicitly mentioned.*

We recommend the reader to go over the following two definitions carefully and remember them well, as they will be used often. Particularly the definition of an induced subhypergraph might be something slightly different from what one would expect as a generalisation of the definition of an induced subgraph in a graph.

**Definition 8.2** (Independent set). *Given a hypergraph $H = (V, \mathcal{E})$, a subset $U \subseteq V$ is called an* independent set *in $H$ if it does not contain any hyperedge of cardinality at least 2, i.e. for every $S \in \mathcal{E}$ with $|S| \geq 2$, we have $S \not\subseteq U$.*

**Definition 8.3** (Subhypergraph $H[V']$). *Let $H = (V, \mathcal{E})$ be a hypergraph. For a subset $V' \subseteq V$, we refer to $H[V'] := (V', \{S \cap V' \mid S \in \mathcal{E}\})$ as the* subhypergraph *of $H$ induced by $V'$.*

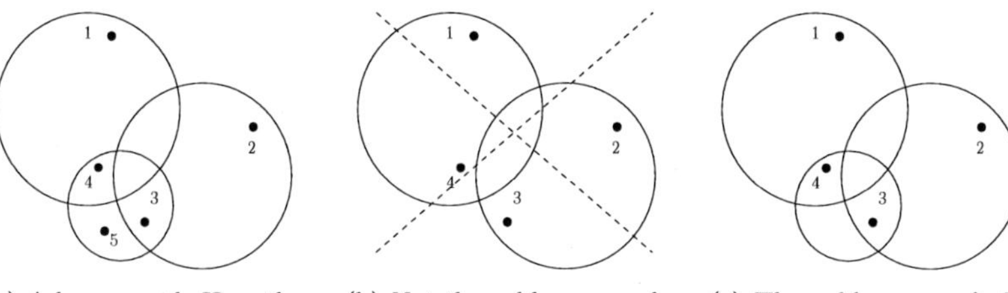

(a) A hypergraph $H$ on the vertex set $\{1, 2, 3, 4, 5\}$

(b) Not the subhypergraph induced by $\{1, 2, 3, 4\}$!

(c) The subhypergraph $H'$ induced by $\{1, 2, 3, 4\}$

Figure 8.1: Induced subhypergraph: the right and the wrong

Many would expect the edge set to be $\mathcal{E} \cap 2^{V'}$ but this is not the case. See Figure 8.1 for a clarifying example.

### 8.1.2 Chromatic numbers

Now we are ready to introduce the colourings and chromatic numbers for hypergraphs. All of these coincide with the usual notion of graph colouring for hypergraphs with all edges of size two.

**Definition 8.4** (Colourings: conflict-free and unique-max). *Let $H = (V, \mathcal{E})$ be a hypergraph and let $C$ be a colouring $C\colon V \to \mathbb{N}$:*

- *We say that $C$ is a* proper colouring *if for every hyperedge $S \in \mathcal{E}$ with $|S| \geq 2$ there exist two vertices $u, v \in S$ such that $C(u) \neq C(v)$. That is, no hyperedge with at least two vertices is monochromatic.*

- *We say that $C$ is a* conflict-free colouring *(cf-colouring in short) if for every hyperedge $S \in \mathcal{E}$ there exists a colour $i \in \mathbb{N}$ such that $|S \cap C^{-1}(i)| = 1$. That is, every hyperedge $S \in \mathcal{E}$ contains some vertex whose colour is unique in $S$.*

- *We say that $C$ is a* unique-maximum colouring *(um-colouring in short) if for every hyperedge $S \in \mathcal{E}$, $|S \cap C^{-1}(\max_{v \in S} C(v))| = 1$. That is, in every hyperedge $S \in \mathcal{E}$ the maximum colour in $S$ is unique in $S$.*

- *We say that $C$ is a* rainbow colouring *(rb-colouring in short) if for every hyperedge $S \in \mathcal{E}$, $|C(S)| = |S|$. That is, in every hyperedge $S \in \mathcal{E}$ the vertices have distinct colours.*

*The non-empty sets $C^{-1}(i)$ are called* colour classes *of the colouring $C$.*

We denote by $\chi(H)$, $\chi_{\mathrm{cf}}(H)$, $\chi_{\mathrm{um}}(H)$ and $\chi_{\mathrm{rb}}(H)$ the minimum integer $k$ for which $H$ admits a proper, a conflict-free, a unique-maximum and a rainbow colouring, respectively, with a total of $k$ colours and call these the proper (or usual), the conflict-free, the unique-maximum and the rainbow *chromatic number* of $H$, respectively. Obviously, every rb-colouring of $H$ is a um-colouring of $H$, every um-colouring of $H$ is a cf-colouring of $H$ and every cf-colouring of $H$ is a proper colouring of $H$, but the converse is not necessarily true. Thus, we have:

$$\chi(H) \leq \chi_{\mathrm{cf}}(H) \leq \chi_{\mathrm{um}}(H) \leq \chi_{\mathrm{rb}}(H).$$

Observe that all four values $\chi(H), \chi_{\mathrm{cf}}(H), \chi_{\mathrm{um}}(H), \chi_{\mathrm{rb}}(H)$ are finite, since a colouring of $H$ on $n$ vertices by $n$ distinct colours is proper, conflict-free, unique-maximum as well as rainbow. An example of a hypergraph which needs so many colours is the complete graph $K_n$.

**Definition 8.5** (Hereditary colourability). *We say that a hypergraph $H$ is* hereditarily $k$-colourable, *if for every subset $V' \subseteq V$, the subhypergraph $H[V']$ induced by $V'$ admits a proper $k$-colouring.*

The colourings of the subhypergraphs might be completely different. An analogous notion can be defined for the other colourings as well but we do not need it.

As a warm-up in the definitions, let us look at the following example:

**Example 8.6.** *Consider the hypergraph $H = ([6], \binom{[6]}{4})$. Its chromatic numbers are: $\chi(H) = 2, \chi_{\mathrm{cf}}(H) = \chi_{\mathrm{um}}(H) = 4$ and $\chi_{\mathrm{rb}}(H) = 6$.*

*Proof.* The usual and rainbow chromatic numbers are the easiest to determine. Each hyperedge has to contain at least two colours, hence $\chi(H) \geq 2$. On the other hand a 2-colouring which colours three vertices by colour 1 and the remaining three vertices by colour 2 is proper: every edge has cardinality four and thus, has to contain both colours. This implies $\chi(H) = 2$.

One can easily see that $\chi_{\mathrm{rb}}(H) \geq 6$ because every two vertices lie in a common hyperedge and hence, have distinct colours in every rb-colouring. Since $H$ has 6 vertices, $\chi_{\mathrm{rb}}(H) = 6$.

For cf and um chromatic number, the situation is slightly more complicated. Observe that at most one colour class has size greater than one. Indeed, assume to the contrary that colour $a$ appears at vertices $u, v$ and colour $b$ at vertices $x, y$. Then the edge $\{u, v, x, y\}$ does not have a unique colour and the colouring is not conflict-free. It is easy to see that each colour class has size at most three as otherwise it would contain an edge. Combining the above two observations yields a lower bound $4 \leq \chi_{\mathrm{cf}}(H) \leq \chi_{\mathrm{um}}(H)$. The colouring $C\colon \{1,2,3\} \to 1, \{4\} \to 2, \{5\} \to 3, \{6\} \to 4$ is a 4-colouring of $H$ and it is not hard to see that it has the unique-maximum property. Hence, $\chi_{\mathrm{cf}}(H) = \chi_{\mathrm{um}}(H) = 4$. □

One important class of hypergraphs, for which cf- and um-colourability has been investigated are hypergraphs arising from connectivity properties of graphs.

**Definition 8.7** (Path hypergraph $H_G^{\mathrm{path}}$). *Given a simple graph $G = (V, E)$, consider the hypergraph*

$$H_G^{\mathrm{path}} := (V, \{S \mid S \text{ is the vertex set of a simple path in } G\}).$$

*A cf- (respectively, um-) colouring of $H_G^{\mathrm{path}}$ is called a cf- (respectively, um-) colouring of $G$* with respect to paths.

Unique-maximum colouring of a graph $G$ with respect to paths is known in the literature as *vertex ranking* or *ordered colouring* (see, e.g. Deogun et al. [DKKM94] or Katchalski et al. [KMS95]).

### 8.1.3 Geometric hypergraphs

Initially, the research of conflict-free colourings was motivated by its application to frequency assignment in cellular networks, where many geometrically defined hypergraphs arise naturally. These were also among the most studied classes of hypergraphs in the context of conflict-free colouring. Here we introduce two common notions of geometric hypergraphs and give several examples.

**Definition 8.8** (Geometric hypergraphs: points and regions – $H_{\mathcal{R}}(P)$). *Let $P$ be a set of $n$ points in the plane and let $\mathcal{R}$ be a family of regions in the plane (such as all discs, all axis-parallel rectangles, etc.). We denote by $H = H_{\mathcal{R}}(P)$ the hypergraph on the set $P$ whose hyperedges are all subsets $P'$ that can be cut out of $P$ by a region in $\mathcal{R}$. That is, $H = (P, \{P \cap R\colon R \in \mathcal{R}\})$. We refer to such a hypergraph as the hypergraph* induced by $P$ with respect to $\mathcal{R}$.

**Definition 8.9** (Geometric hypergraphs: regions – $H(\mathcal{R})$). *For a finite family $\mathcal{R}$ of planar regions, we denote by $H(\mathcal{R})$ the hypergraph whose vertex set is $\mathcal{R}$ and whose hyperedge set is the family $\{\mathcal{R}_p \mid p \in \mathbb{R}^2\}$ where $\mathcal{R}_p \subseteq \mathcal{R}$ is the subset of all regions in $\mathcal{R}$ that contain $p$. We refer to such a hypergraph as the hypergraph* induced by $\mathcal{R}$.

We introduce notation for some of the most prominent families of regions:

- $\mathcal{I}$ denotes the family of all closed intervals on the real line,
- $\mathcal{H}$ denotes the family of all closed half-planes in the plane,
- $\mathcal{D}$ denotes the family of all closed discs in the plane, and
- $\mathcal{A}$ denotes the family of all closed axis-parallel rectangles in the plane.

Probably the simplest geometric hypergraph one can think of is a hypergraph induced by points with respect to intervals. It is not hard to see that the actual point set does not matter and any $n$ points $P$ result in the same hypergraph $H_{\mathcal{I}}(P)$ (up to an isomorphism). The *discrete interval hypergraph* on $n$ vertices is the hypergraph $H_n := H_{\mathcal{I}}([n])$, i.e. the hypergraph with vertex set $[n]$ and hyperedge set $\{[s,t] \mid s \leq t,\ s,t \in [n]\}$.

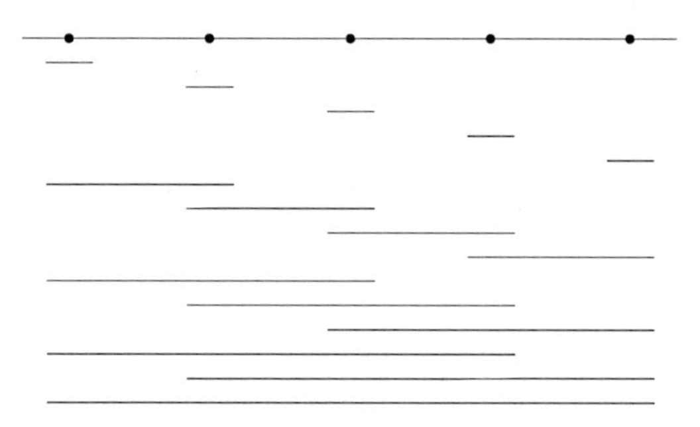

Figure 8.2: Example of a discrete interval hypergraph: on the top are the points on the line and in the lower part the hyperedges of $H_5$.

Let us investigate the conflict-free chromatic number of $H_n$ as another warm-up exercise. We return to this exercise later, in Section 9.1, in the context of its cf and um-choosability.

**Proposition 8.10.** *For every $n \geq 1$, $\chi_{\mathrm{cf}}(H_n) \geq \lfloor \log_2 n \rfloor + 1$.*

*Proof.* Assume, without loss of generality, that $n = 2^{k+1} - 1$. We will show that any cf-colouring of $H_n$ uses at least $k+1$ colours. We proceed by induction on $k$. Consider an arbitrary cf-colouring $C : [n] \to [\ell]$. By the conflict-free property, there is a vertex $i \in [n]$ whose colour $C(i)$ is unique in the whole interval $[n]$. Without loss of generality, $i > 2^k - 1$. Any hyperedge $[a,b]$ with $a \leq i \leq b$ contains a unique colour $C(i)$ but all the hyperedges $[a,b]$ with $a \leq b \leq i-1$ have to contain another unique colour. Hence, we require that the colouring of the hypergraph $([i-1], \{[a,b] \colon a,b \in \mathbb{N}, 1 \leq a \leq b \leq i-1\})$

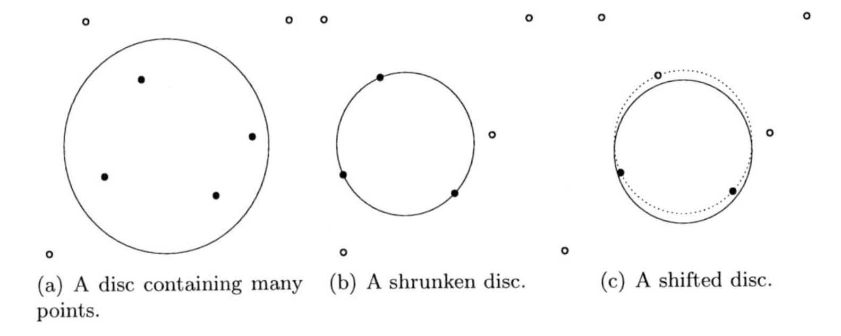

(a) A disc containing many points. (b) A shrunken disc. (c) A shifted disc.

Figure 8.3: Shrinking discs

is again conflict-free. A brief meditation reveals that we have just written the hypergraph $H_{i-1}$ which requires at least $k$ colours by the inductive hypothesis. Since $C(i)$ is globally unique and $i \notin [2^k - 1]$, the colouring $C$ uses at least $k+1$ colours. □

The above bound is tight, and indeed $\chi_{\mathrm{cf}}(H_n) \leq \lfloor \log_2 n \rfloor + 1$. The proof of the matching upper bound is not difficult but we postpone it to the beginning of Section 9.1 where we prove a more general statement. We encourage the impatient readers to try to prove the upper bound themselves, for the time being.

Studying the (proper) colourings of hypergraphs can get quite complicated even for small examples. However, for some geometric hypergraphs there is a tool which makes this task much easier.

**Definition 8.11.** *Let $H = (V, \mathcal{E})$ be a hypergraph. Let $G(H) = (V, E)$ denote the graph whose edges are all hyperedges of $\mathcal{E}$ with cardinality two. We refer to $G$ as the* Delaunay graph *of $H$.*

Consider a hypergraph $H_{\mathcal{D}}(P)$ induced by a set $P$ of $n$ points in the plane with respect to discs. This hypergraph has a very useful property: For every hyperedge $S \in \mathcal{E}(H_{\mathcal{D}}(P))$ with $|S| > 2$, there is a hyperedge $S' \in \mathcal{E}(H_{\mathcal{D}}(P))$ satisfying $S' \subseteq S$ and $|S'| = 2$. To see why, consider a disc $D \in \mathcal{D}$ such that $S = D \cap P$. We can shrink the disc concentrically until the interior of the disc contains one or no points for the first time (see Figure 8.3(b)). Denote $D'$ the resulting disc. If $|D' \cap P| > 2$, then we have at least two points of $P$ on the boundary of $D'$ (as otherwise, we could continue shrinking it concentrically). Moving the disc $D'$ away from one of these points, say $p$, by a sufficiently small distance results in another disc $D''$ which has $D'' \cap P = (D' \cap P) \setminus p$

(see Figure 8.3(c) on the preceding page). Iterating the two actions at most $n$ times results in a disc $D^*$ with $|D^* \cap P| = 2$ as required.

This property is common to many hypergraphs induced by points with respect to some family of regions.

**Definition 8.12** (Shrinkability). *Let $H = (V, \mathcal{E})$ be a hypergraph. We call $H$* shrinkable, *if every hyperedge $S \in \mathcal{E}$ contains a hyperedge $S' \in \mathcal{E}, S' \subseteq S$ of cardinality $|S'| = 2$.*

Colourings of shrinkable hypergraphs have a simpler structure.

**Observation 8.13** (Colouring Delaunay graph). *Let $H = (V, \mathcal{E})$ be a hypergraph, then the following statements hold:*

(i) *any proper colouring of the hypergraph $H$ is also a proper colouring of its Delaunay graph $G(H)$,*
(ii) *if $H$ is shrinkable, then also every proper colouring of $G(H)$ is a proper colouring of the hypergraph $H$,*
(iii) *if $H$ is shrinkable, then $\chi(H) = \chi(G(H))$.*

*Proof.* (i) This implication is trivial, since $E \subseteq \mathcal{E}$.
(ii) Let $S \in \mathcal{E}$ be a hyperedge. By the shrinkability, there is an edge $e \in E$ with $e \subseteq S$. It must be non-monochromatically coloured since it is an edge of $G$. Consequently, $S$ is non-monochromatic because $e \subseteq S$.
(iii) Follows from (i) and (ii).

□

Another important characteristic of a family $\mathcal{R}$ of regions, which has a huge impact on the colourability of the underlying geometric hypergraphs, is the union complexity of $\mathcal{R}$.

**Definition 8.14** (Union complexity). *Let $\mathcal{R}$ be a family of $n$ simple Jordan regions in the plane. The* union complexity $\mathcal{U}(\mathcal{R})$ *of $\mathcal{R}$ is the number of vertices (i.e. intersections of boundaries) lying on the boundary $\partial (\cup \mathcal{R})$ of the union of the regions.*

We will see later (in Theorem 8.29) that low union complexity in a hereditary sense implies colourability with few colours.

## 8.2 Motivation

The investigation of cf- and um-colouring [ELRS03, Smo03] was motivated by an application of conflict-free colourings to the frequency assignment in cellular networks. An independent application of unique-maximum colourings is the use of vertex ranking for scheduling certain parallel tasks.

### 8.2.1 Frequency assignment

Wireless communication plays a crucial role in our lives, nowadays, and we encounter it on a daily basis: whenever we use our mobile phone or connect our laptop to a wireless access point. But usually, we know little about interesting mathematical problems that lurk behind.

One of the crucial tasks in the background of the wireless communication is assigning the frequencies to the connections (between access points and laptops or between base stations and mobile phones) while avoiding interferences between them. An interference would typically occur between two geographically close connections which are assigned similar frequencies (or channels). Rapid development of wireless networks led to scarcity of frequencies and assigning their limited range became a non-trivial goal to achieve. This situation has been first modelled and mathematically formalised in the 1960's by Metzger [Met70] and many variants have been studied since (for an overview of optimisation methods and heuristics used in several static variants of the problem, see the surveys of Aardal et al. [AvHK+03, AvHK+07]).

The frequency assignment refers to a variety of mathematical models and problems related to the scenario described above. We only review one particular variant of the problem for cellular networks from Even et al. [ELRS03], to which the conflict-free colourings are related.

A cellular network consists of two kinds of nodes: *base stations* and *mobile clients*. The base stations have fixed positions and are interconnected by a backbone network, whereas the clients can move around and can only connect to the base stations by radio links. Each base station gets a fixed frequency assigned and only communicates with the clients on this frequency. The clients switch their frequencies as they move, depending on which frequency they can use at a given location. Base stations with the same frequency interfere and the clients cannot communicate with either of them, if they are in the range of several. The key problem is to assign frequencies to the base stations in such a way that every client communicates with some base station.

Let us formulate this in the language of hypergraph colouring. Let $\mathcal{D}' \subseteq \mathcal{D}$ be the set of discs representing the antennae. We seek the minimum number of colours $k$ such that one can assign each disc one of the $k$ colours so that at every point $x$ in the union of the discs in $\mathcal{D}'$, there is at least one disc $D \in \mathcal{D}'$ that covers $x$ and whose colour is distinct from all the colours of other discs containing $x$. This is equivalent to finding the cf-chromatic number of the hypergraph $H(\mathcal{D}')$.

### 8.2.2 Vertex ranking

Vertex ranking is a special case of um-colouring and has applications in several areas, including VLSI design (see Leiserson [Lei80] or Sen et al. [SDG92]), parallel sparse Cholesky factorisation (see Liu [Liu86]) and scheduling problems of assembly steps in manufacturing (see Iyer et al. [IRV88]). Generally speaking, vertex ranking often appears in situations when a multitude of tasks with some dependencies needs to be performed efficiently in parallel.

## 8.3 Overview

The study of cf-colouring was initiated by Even et al. [ELRS03] and Smorodinsky [Smo03] and was further studied in many settings.

The original colouring problem was posed for discs in the plane and researchers continued investigating the conflict-free chromatic number of hypergraphs defined by geometric shapes, such as intervals, discs, axis-parallel rectangles, homothetic copies of a convex body, and geometric shapes with small union complexity (see, e.g. [AEGR07, AS08, CPST09, ELRS03, HPS03, HPS05, PT03, Smo03, Smo06, Smo07]). Some effort has been put into finding approximation algorithms for conflict-free colouring discs, axis-parallel rectangles and axis-parallel regular hexagons (see [ELRS03, LTP09]) and several works looked into either weaker (see [HPS03, HPS05, Smo03]) or stronger (see [ABG$^+$05, HKS10]) variants of conflict-freeness in the geometric setting.

It turns out that conflict-free colouring of geometric hypergraphs gets very difficult even for the simplest shapes such as intervals, when one switches to the online setting. This has been investigated under several different online models and for many different geometric hypergraphs defined by intervals, half-planes, unit discs and nearly-equal rectangles to name a few (see, e.g. [BNCOS07a, BNCOS07b, BNCOS10, BNCS06, BNCS08, CCHT08, CFK$^+$07, CKS09, FLM$^+$05, Smo08]).

Most algorithms for conflict-free colouring in the literature produce unique-maximum colourings which have more structure and thus, seem to be easier to argue about in the proofs. An interesting question is how can a non-unique-maximum conflict-free colouring improve on a unique-maximum colouring, with respect to the number of colours used, the research of which has been pursued in [CKP10, CT10].

In the rest of the chapter, we overview some of the most important results for conflict-free colourings so that the reader can make his own picture about the subject and place our results in the context.

### 8.3.1 General results

Although cf-colouring has been mostly studied in the geometric setting (geometric hypergraphs) and the graph setting (vertex ranking), there are some interesting results which hold for arbitrary hypergraphs.

Several such results are in the paper of Pach and Tardos [PT09] who proved non-trivial upper bounds on the cf-chromatic number of hypergraphs. One of them, which we generalise later, is the following:

**Theorem 8.15** ([PT09]). *Let $H = (V, \mathcal{E})$ be a hypergraph and $m := |\mathcal{E}|$. Then*

$$\chi_{\mathrm{cf}}(H) \leq \frac{1}{2} + \sqrt{2m + \tfrac{1}{4}}. \tag{8.1}$$

A general framework for conflict-free colouring hypergraphs was introduced by Smorodinsky [Smo06, Smo07]. This framework is, intuitively speaking, based on colouring a large independent set in the hypergraph by one colour and proceeding recursively on the subhypergraph induced by the remaining vertices with the remaining colours. A large independent subset is always guaranteed, when the hypergraph has a small chromatic number.

**Theorem 8.16** ([Smo06, Smo07]). *Let $H$ be a hereditarily $k$-colourable hypergraph. Then*

$$\chi_{\mathrm{cf}}(H) \leq \chi_{\mathrm{um}}(H) \leq \log_{\frac{k}{k-1}} n + 1 \in O(k \log n). \tag{8.2}$$

We will see many interesting consequences of Theorem 8.16 for geometric hypergraphs in the pages that follow.

### 8.3.2 Geometric hypergraphs

The framework described above tells us that a (subhypergraph-closed) family of hypergraphs has its conflict-free chromatic number bounded in its usual chromatic number (and $\log n$). Let us start by reviewing the usual chromatic number of several hypergraphs of interest.

A discrete interval hypergraph $H_n$ is obviously shrinkable and its Delaunay graph is a path which by Observation 8.13 implies:

**Proposition 8.17.** *Let $P$ be a set of $n$ points in $\mathbb{R}$. Then $\chi(H_{\mathcal{I}}(P)) \leq 2$.*

A hypergraph $H$ induced by a set of points $P$ with respect to half-planes is also shrinkable and hence $\chi(H) = \chi(G(H))$. The Delaunay graph $G(H)$ consists of exactly all the 2-sets of the point set $P$. These are of two types: (i) convex hull edges of $P$ and (ii) sets $\{u, v\}$ where exactly one of the points is a convex hull vertex. We can colour the convex hull vertices by three colours, since they form a cycle in $G(H)$. If one of the remaining points $p$ lies in three or more 2-sets, then there are three half-planes $h_1, h_2, h_3$ such that $h_1 \cup h_2 \cup h_3 = \mathbb{R}^2$ and each $h_i$ contains $p$ and exactly one vertex of the convex hull. Thus, $P$ consists of four vertices in non-convex position. Otherwise, each other point is incident to at most two convex hull vertices and hence, can be coloured by a remaining colour which proves:

**Proposition 8.18.** *Let $P$ be a set of $n$ points in the plane other than four points in non-convex position. Then $\chi(H_{\mathcal{H}}(P)) \leq 3$.*

A hypergraph induced by points with respect to discs is shrinkable too, as we have seen in Section 8.1.3. Hence, it has the same chromatic number as its Delaunay graph which is a Delaunay triangulation (see, e.g. [dBCvKO08]). The triangulation is planar and hence, 4-colourable by the 4-colour theorem (see, e.g. [AH89, RSST96]) which implies:

**Proposition 8.19.** *Let $P$ be a set of $n$ points in the plane. Then $\chi(H_{\mathcal{D}}(P)) \leq 4$.*

For hypergraphs induced by families of discs, Smorodinsky [Smo06, Smo07] proved the following:

**Proposition 8.20** ([Smo06, Smo07]). *Let $\mathcal{D}' \subseteq \mathcal{D}$ be a family of $n$ discs in the plane. Then $\chi(H(\mathcal{D}')) \leq 4$.*

The framework of Theorem 8.16 immediately implies:

**Theorem 8.21** ([ELRS03, Smo03]). *Let $P$ be a set of $n$ points in $\mathbb{R}$. Then $\chi_{\text{cf}}(H_{\mathcal{I}}(P)) \leq \log_2 n + 1$ (this is a matching bound to Proposition 8.10 on page 110).*

**Theorem 8.22.** *Let $P$ be a set of $n$ points in the plane. Then we have $\chi_{\text{cf}}(H_{\mathcal{H}}(P)) \leq \log_{3/2} n + 1$.*

**Theorem 8.23** ([ELRS03, Smo03]). *Let $P$ be a set of $n$ points in the plane. Then $\chi_{\text{cf}}(H_{\mathcal{D}}(P)) \leq \log_{4/3} n + 1$.*

**Theorem 8.24** ([Smo06, Smo07]). *Let $\mathcal{D}' \subseteq \mathcal{D}$ be a family of $n$ discs in the plane. Then $\chi_{\text{cf}}(H(\mathcal{D}')) \leq \log_{4/3} n + 1$.*

Another natural class of shapes to consider are rectangles. As we will shortly see, it is more interesting to restrict ourselves to axis-parallel rectangles.

There are hypergraphs induced by a family $\mathcal{R}$ of $n$ arbitrary rectangles which contain the complete graph $K_n$ as a subhypergraph and hence, cannot be cf-coloured by fewer than $n$ colours. An example of such family $\mathcal{R}$ is the following: Consider $n$ lines $\ell_1, \ldots, \ell_n$ in the plane in general position (no two are parallel and no three intersect in a common point). Take a sufficiently large disc $D$ with all the intersections points of the lines inside $D$ and define segments $s_1, \ldots s_n$ as $s_i := \ell_i \cap D$. It holds true that every two of the segments intersect and their intersection point does not lie on any other segment. To finish the argument, one can either think of segments as a special case of rectangles, or "thicken" them to a sufficiently small non-zero width to obtain the final family $\mathcal{R}$.

A hypergraph induced by a set $P$ of $n$ points in general position with respect to segments must also necessarily contain all edges of the complete graph $K_n$ as hyperedges, since a pair of points is contained in a segment containing only that pair. As such, it cannot be coloured with fewer than $n$ colours.

For axis-parallel rectangles, Smorodinsky [Smo06, Smo07] proved:

**Theorem 8.25** ([Smo06, Smo07]). *Let $\mathcal{A}' \subseteq \mathcal{A}$ be a family of $n$ axis-parallel rectangles in the plane. Then $\chi(H(\mathcal{A}')) \leq 8(\log_2 n + 1)$.*

Moreover, Pach and Tardos [PT10] proved that this bound is asymptotically tight, i.e. there are families $\mathcal{A}_n$ of axis-parallel rectangles which have chromatic number $\chi(H(\mathcal{A}_n)) \in \Omega(\log n)$. The framework from Theorem 8.16 again implies a result for the cf-colouring immediately:

**Theorem 8.26.** *Let $\mathcal{A}' \subseteq \mathcal{A}$ be a family of $n$ axis-parallel rectangles in the plane. Then $\chi_{\mathrm{cf}}(H(\mathcal{A}')) \in O(\log^2 n)$.*

The logarithmic lower bound on the usual chromatic number mentioned above immediately implies a logarithmic lower bound on the cf-chromatic number as well.

For a hypergraph $H_{\mathcal{A}}(P)$ induced by a set $P$ of $n$ points with respect to axis-parallel rectangles, Har-Peled and Smorodinsky [HPS03, HPS05] observed that $\chi(H_{\mathcal{A}}(P)) \leq \sqrt{n}$ which implies $\chi_{\mathrm{cf}}(H_{\mathcal{A}}(P)) \leq \sqrt{n} \log n$. A recent improvement by Ajwani et al. [AEGR07] yields:

**Theorem 8.27** ([AEGR07]). *Let $P$ be a set of $n$ points in the plane. Then we have $\chi(H_{\mathcal{A}}(P)) \in O(n^{0.382})$ and $\chi_{\mathrm{cf}}(H_{\mathcal{A}}(P)) \in O(n^{0.382})$.* *

*The exact exponent proved for both chromatic numbers is $\frac{3-\sqrt{5}}{2}$ and additional multiplicative polylogarithmic factors are involved.

The currently best lower bound is due to Chen et al. [CPST09] and the gap is still wide open.

**Theorem 8.28** ([CPST09]). *There are n-element point sets $P_n$ in the plane with chromatic number* $\chi(H_{\mathcal{A}}(P_n)) \in \Omega\left(\frac{\log n}{\log^2 \log n}\right)$.

A general tool for bounding chromatic number of geometric hypergraphs (and consequently, also often bounding cf-chromatic number) was developed by Smorodinsky [Smo07]:

**Theorem 8.29** ([Smo07]). *Let $\mathcal{R}$ be a set of $n$ simple Jordan regions. If $\mathcal{U} : \mathbb{Z}_0^+ \to \mathbb{Z}_0^+$ is a non-decreasing function such that $\mathcal{U}(m)$ upper bounds the union complexity of any $k \leq m$ of the regions, then* $\chi(H(\mathcal{R})) \in O\left(\frac{\mathcal{U}(n)}{n}\right)$.

A prime example of regions with small union complexity are discs and pseudo-discs: for a family $\mathcal{P}$ of $n$ pseudo-discs, the union complexity is always bounded by $\mathcal{U}(\mathcal{P}) \leq 6|\mathcal{P}|$ (see, e.g. Kedem et al. [KLPS86]). As consequences, we obtain the following:

**Theorem 8.30.** *Let $\mathcal{P}$ be a family of $n$ pseudo-discs. Then* $\chi(H(\mathcal{P})) \in O(1)$.

**Theorem 8.31.** *Let $\mathcal{P}$ be a family of $n$ pseudo-discs. Then* $\chi_{\mathrm{cf}}(H(\mathcal{P})) \in O(\log n)$.

A very wide variety of other families of planar regions have near-linear union complexity. We include an overview in Table 8.1 (the families are $t$-intersecting). The definitions of these classes as well as the precise statements of the upper bounds are included in Appendix A.

| family | union complexity | reference |
|---|---|---|
| $\Delta$-fat triangles | $O(n2^{\alpha(n)} \log^* n)$ | Aronov et al. [EAS11] |
| locally $\gamma$-fat objects | $O(\lambda_{t+2}(n) \log n)$ | de Berg [dB10] |

Table 8.1: Families of planar objects with low union complexity.

### 8.3.3 Variations

Several variations of the cf-colourings have been studied. The most notable ones are $k$-cf-colourings, $k$-strong cf-colourings.

**$k$-cf-colourings.** One possible relaxation of the conflict-free property is the following: instead of insisting on hyperedges containing a unique colour, we will be satisfied if every hyperedge $S$ contains a colour which appears at most $k$ times in $S$. Such a colouring is called a *$k$-cf-colouring*. Obviously, 1-cf-colourings are exactly cf-colourings. This variant has mainly been investigated for balls in $\mathbb{R}^3$ or for large values of $k$ (see Smorodinsky [Smo03] and Har-Peled and Smorodinsky [HPS03, HPS05]).

**$k$-strong cf-colourings.** One can similarly strengthen the conflict-free property: instead of asking for one unique colour, we require $\min\{k, |S|\}$ unique colours in a hyperedge $S$. Such a colouring is called a *$k$-strong cf-colouring*. Again, 1-strong cf-colourings are exactly cf-colourings. These have been looked into in the disc setting by Abellanas et al. [ABG$^+$05] and a general framework relating them to another class of $k$-colourful colourings was found by Horev et al. [HKS10].

### 8.3.4 Online algorithms

Until now, we only considered a *static model* of cf-colouring of hypergraphs, where the whole hypergraph $H$ is given and a conflict-free colouring must be found efficiently.

Bar-Noy et al. [BNCS08] introduced a hierarchy of *dynamic models*, where a sequence $(H^t)_{t=0}^n$ of hypergraphs is given, with $H^t$ on $t$ vertices and additionally, $H^{t-1}$ is an induced subhypergraph of $H^t$. At discrete times $t$, the hypergraph $H^t$ is revealed and a conflict-free colouring $C^t : V(H^t) \to \mathbb{N}$ of $H^t$ extending the previous colouring $C^{t-1}$ (i.e. $C^t|_{V(H^{t-1})} = C^{t-1}$) has to be constructed.

Let us illustrate this on an example of geometric hypergraphs induced by a set of $n$ points $P$ in the plane with respect to some regions $\mathcal{R}$. Order the points in $P$ in a sequence $(p_1, \ldots, p_n)$ and denote $P^t := \{p_1, \ldots, p_t\}$ the set of points revealed until time $t$. Consider the hypergraph sequence $H^t := H_{\mathcal{R}}(P^t)$. It is easy to verify that $H^t[P^{t-1}] = H^{t-1}$, since $(R \cap P^t) \cap P^{t-1} = R \cap P^{t-1}$ for any region $R \in \mathcal{R}$ as required. Any sequence of points defines a corresponding sequence of hypergraphs and we will refer to the corresponding dynamic hypergraph colouring problem as to *dynamic colouring points with respect to $\mathcal{R}$*. The dynamic colouring has several models:

In the *dynamic offline model*, the complete sequence $(H^t)_{t=1}^n$ is given. Notice that this model is conceptually the same as static colouring of the hypergraph $H := (V(H^n), \bigcup_{t\in[n]} \mathcal{E}(H^t))$.

In the dynamic online models, the sequence $(H^t)_{t=1}^n$ is revealed incrementally and a colour of the new vertex $v^t$ has to be assigned without the knowledge of the future $(H^s)_{s>t}$.

In the *dynamic online-with-absolute-positions model*, the final hypergraph $H^n$ is labelled and known from the beginning. The hypergraphs $H^t$ are revealed together with the vertex labels, i.e. we know which vertex in $H^t$ corresponds to which vertex in $H^n$.

In the *dynamic online-with-relative-positions model*, no information about the final hypergraph $H^n$ is given (even the size $n$ is not known). The only information we might have is the structure of the hypergraphs (e.g. we know we are colouring points with respect to intervals).

For randomised algorithms, there is a weaker notion of an *oblivious adversary* who knows the colouring algorithm but has to commit to the whole input sequence beforehand, without seeing the random bits.

The dynamic variants, especially the online-with-relative-positions model, turn out to be much more difficult. This is especially staggering for the hypergraph classes, where the usual cf-colouring is very easy, such as points with respect to intervals. The known bound are:[1]

### Dynamic offline model

In the dynamic offline model, Bar-Noy et al. [BNCS08] observed that points $P$ in $\mathbb{R}$ with respect to intervals have the following property:

**Lemma 8.32** ([BNCS08]). *Let $(x_1, \ldots, x_n)$ be a sequence of points in $\mathbb{R}$ and denote $P^t := \{x_1, \ldots, x_t\}$. Then the hypergraph $H := (P, \cup_{t\in[n]} \mathcal{E}(H_{\mathcal{I}}(P^t)))$ is hereditarily 3-colourable.*

This, together with the Theorem 8.16 implies an upper bound of $\log_{3/2} n+1$ colours. On the other hand, they showed that $3 \log_5 n+1$ colours are necessary in the worst case.

For points with respect to discs, there is an insertion sequence of points which produces all the hyperedges of size two [CFK$^+$07] and hence, $n$ colours are necessary. With unit discs, however, $O(\log n)$ colours suffice [CCHT08]. This immediately implies the same bound for points with respect to half-planes.

[1]Note that a lower bound in a more restricted model implies the same lower bound on less restricted model and the opposite holds for upper bounds.

### Online absolute-positions model

Bar-Noy et al. [BNCS06, BNCS08] gave an algorithm for colouring $n$ points with respect to intervals using $3(\log_3 n + 1)$ colours.

### Online relative-positions model

An algorithm for online colouring $n$ points with respect to intervals with $O(\log^2 n)$ colours was found in [CFK$^+$07, FLM$^+$05]. There is still no algorithm known to achieve $O(\log n)$ colours (which is the lower bound for the static cf-colouring and hence, the best we can hope for). However, with a randomised algorithm against an oblivious adversary, $O(\log n)$ colours suffice with high probability.[¬] The same holds for $n$ points in the plane with respect to unit discs and $n$ points in the plane with respect to half-planes (see, e.g. [CKS09], or [BNCOS07b, BNCOS10] for a general online cf-colouring framework).

## 8.4 List colouring

Until now, research on cf-colouring was carried out under the assumption that we can use any colour from some global set of colours.[§] The goal was to minimise the total number of colours used. In real life, it makes sense to assume that each antenna in the wireless network is further restricted to use a subset of the available spectrum. This restriction might be local (depending, say, on the physical location of the antenna). Hence, different antennae may have different subsets of (admissible) frequencies available for them. That is, assume further that each antenna $D \in \mathcal{D}'$ is associated with a subset $L_D$ of frequencies. We want to assign to each antenna $D$ a frequency that is taken from its allowed set $L_D$. The following problem arises: *What is the minimum number $f = f(n)$ such that given any set $\mathcal{D}'$ of $n$ antennae (represented as discs) and any family of subsets of positive integers $\mathcal{L} = \{L_D\}_{D \in \mathcal{D}'}$ associated with the antennae in $\mathcal{D}'$, the following holds: If each subset $L_D$ is of cardinality $f$, then one can cf-colour the hypergraph $H = H(\mathcal{D}')$ from $\mathcal{L}$.* In what follows, we give a formal definition of the colouring model.

**Definition 8.33.** *Let $H = (V, \mathcal{E})$ be a hypergraph and let $\mathcal{L} = \{L_v\}_{v \in V}$ be a family of $|V|$ subsets of $\mathbb{N}$. We say that $H$* admits a cf-colouring from $\mathcal{L}$ *if there exists a cf-colouring $C\colon V \to \mathbb{N}$ such that $C(v) \in L_v$ for every $v \in V$. Analogous definitions apply for the notions of a hypergraph $H$* admitting a proper, a um- *and* a rb-colouring from $\mathcal{L}$.

[¬]Note that these results still imply the same bounds in the dynamic offline model.
[§]With the exception of a recent preprint of Cheilaris and Smorodinsky [CS10]

**Definition 8.34.** *We say that a hypergraph $H = (V, \mathcal{E})$ is* $k$-cf-choosable *if for every family $\mathcal{L} = \{L_v\}_{v \in V}$ such that $|L_v| \geq k$ for all $v \in V$, $H$ admits a cf-colouring from $\mathcal{L}$. Analogous definitions apply for the notions of a hypergraph $H$ being* $k$-choosable, $k$-um-choosable *and* $k$-rb-choosable.

In this work, we are interested in the minimum number $k$ for which a given geometric hypergraph is $k$-cf-choosable (respectively, $k$-choosable, $k$-um-choosable, and $k$-rb-choosable). We refer to this number as the *cf-choice number* (respectively, *choice number*, *um-choice number* and *rb-choice number*) of $H$ and denote it by $ch_{\mathrm{cf}}(H)$ (respectively, $ch(H)$, $ch_{\mathrm{um}}(H)$ and $ch_{\mathrm{rb}}(H)$). Obviously, if the cf-choice number of $H$ is $k$ then $\chi_{\mathrm{cf}}(H) \leq k$, as one can cf-colour $H$ from $\mathcal{L} = \{L_v\}_{v \in V}$ where for every $v$ we have $L_v = \{1, \ldots, k\}$ (the same can be said for proper, um- and rb-colourings). Thus,

$$\begin{array}{ll} ch(H) \geq \chi(H), & ch_{\mathrm{cf}}(H) \geq \chi_{\mathrm{cf}}(H), \\ ch_{\mathrm{um}}(H) \geq \chi_{\mathrm{um}}(H), & ch_{\mathrm{rb}}(H) \geq \chi_{\mathrm{rb}}(H). \end{array} \tag{8.3}$$

It is also easy to see that all of those parameters are upper-bounded by the number of vertices of the underlying hypergraph.

The study of list colouring for the special case of graphs, i.e. 2-uniform hypergraphs, was initiated in [ERT80, Viz76]. List proper colouring of hypergraphs has been studied more recently, as well; see, e.g. [KV01]. We refer the reader to the survey of Alon [Alo93] for more on list colouring of graphs.

# 9

# Potential method

In this chapter we study the cf-choice number, and the um-choice number of geometric hypergraphs.

Our main result is a generalisation of the framework from Theorem 8.16 on page 115 to list um-colouring. It turns out that list um-colouring is non-trivial even for points with respect to intervals and a new approach is necessary.

In Section 9.1, we introduce a general algorithm for list um-colouring hypergraphs with a potential-based rule for choosing the right vertices to colour. The algorithm allows us to colour an arbitrary hypergraph from lists of possibly different sizes, under some condition on the list sizes. We show that the condition is in a sense best possible. Surprisingly, we do not lose anything and the potential method yields exactly the same bounds for um-choice number as Theorem 8.16 does for cf and um-chromatic number.

In Section 9.2, we apply the potential method to list um-colouring several geometric hypergraphs and obtain corollaries analogous to those of Theorem 8.16 for the cf-chromatic number. We give an $O(\log n)$ bound on the um-choice number of hypergraphs induced by points with respect to intervals, discs and half-planes as well as hypergraphs induced by discs or pseudo-discs and more generally, families of regions with linear (hereditary) union complexity

This is joint work with Panos Cheilaris and Shakhar Smorodinsky [CSS10].

## 9.1 A potential method for list colouring

Let us return to the simple example of the discrete interval hypergraph $H_n$ (i.e. $H_{\mathcal{I}}([n])$). In Proposition 8.10 we have seen that $\chi_{\text{cf}}(H_n) \geq \lfloor \log_2 n \rfloor + 1$ and promised to return to this example and prove an upper bound in the choosability setting. Since $ch_{\text{cf}}(H) \geq \chi_{\text{cf}}(H)$ for every hypergraph $H$, we have $ch_{\text{cf}}(H_n) \geq \lfloor \log_2 n \rfloor + 1$. As a preparation for the rest of this section, we prove that this bound is tight:

**Proposition 9.1.** *For every $n \geq 1$, $ch_{\text{cf}}(H_n) \leq \lfloor \log_2 n \rfloor + 1$.*

*Proof.* Assume, without loss of generality, that $n = 2^{k+1} - 1$. We will show that $H_n$ is $k+1$ cf-choosable. The proof is by induction on $k$. Let $\mathcal{L} = \{L_i\}_{i \in [n]}$, such that $|L_i| = k+1$, for every $i$. Consider the median vertex $p = 2^k$. Choose a colour $x \in L_p$ and assign it to $p$. Remove $x$ from all the other lists (for lists containing $x$), i.e. consider $\mathcal{L}' = \{L'_i\}_{i \in [n] \setminus \{p\}}$ where $L'_i = L_i \setminus \{x\}$. Note that all the lists in $\mathcal{L}'$ have size at least $k$. The induction hypothesis is that we can cf-colour any set of points of size $2^k - 1$ from lists of size $k$. Indeed, the number of vertices smaller (respectively, larger) than $p$ is exactly $2^k - 1$. Thus, we cf-colour vertices smaller than $p$ and independently vertices larger than $p$, both using colours from the lists of $\mathcal{L}'$. Intervals that contain the median vertex $p$ also have the conflict-free property, because colour $x$ is used only in $p$. This completes the induction step and hence the proof of the proposition. $\square$

We now turn to the more difficult problem of bounding the um-choice number. Even for the discrete interval hypergraph $H_n$, a divide and conquer approach, along the lines of the proof of Proposition 9.1 is doomed to fail. In such an approach, some vertex close to the median must be found, a colour must be assigned to it from its list, and this colour must be deleted from all other lists. However, vertices close to the median might have only 'low' colours in their lists. Thus, while we are guaranteed that a vertex close to the median is uniquely coloured for intervals containing it, such a unique colour is not necessarily the maximal colour for such intervals.

Instead, we use a different approach. It provides a general framework for um-colouring hypergraphs from lists. Moreover, when applied to many geometric hypergraphs, it provides asymptotically tight bounds for the cf and um-choice number.

Below, we give an informal description of the approach which is then summarised in Algorithm 9.1.

We start by sorting the colours in the union of all lists in increasing order. Let $c$ denote the minimum colour and $V^c \subseteq V$ denote the subset of vertices containing $c$ in their lists. Note that $V^c$ might contain very few vertices, in fact, it might be that $|V^c| = 1$. We colour all vertices in a suitable subset $U \subseteq V^c$ with $c$. We make sure that $U$ is independent in the hypergraph $H[V^c]$.* The exact way in which we choose $U$ is crucial to the performance of the algorithm and is discussed below. Next, for the uncoloured vertices in $V^c \setminus U$, we remove the colour $c$ from their lists. This is repeated for every colour in the union $\bigcup_{v \in V} L_v$ in increasing order of the colours. The algorithm stops when all vertices are coloured. Notice that such an algorithm might run into a problem, when all colours in the list of some vertex are removed before this vertex is coloured. Later, we show that if we choose the subset $U \subseteq V^c$ in a clever way and the lists are sufficiently large, then we avoid such a problem.

**Algorithm 9.1** UMColourGeneric($H$, $\mathcal{L}$): Unique-max colour hypergraph $H = (V, \mathcal{E})$ from family $\mathcal{L}$

**while** $V \neq \emptyset$ **do**
  $c \leftarrow \min \bigcup_{v \in V} L_v$ (* *c is the minimum colour among all lists* *)
  $V^c \leftarrow \{v \in V \mid c \in L_v\}$ (* *$V^c$ is the subset of remaining vertices containing c in their lists* *)
  $U \leftarrow$ a "good" independent subset of the induced hypergraph $H[V^c]$
  **for** $x \in U$ **do**
    $f(x) \leftarrow c$ (* *colour it with colour c* *)
  **end for**
  **for** $v \in V^c \setminus U$ **do** (* *for every uncoloured vertex, erase c from its list* *)
    $L_v \leftarrow L_v \setminus \{c\}$
  **end for**
  $V \leftarrow V \setminus U$ (* *remove the coloured vertices* *)
**end while**
**return** $f$

As already mentioned, Algorithm 9.1 might cause some lists to run out of colours before colouring all vertices. However, if this does not happen, we prove that the algorithm produces a um-colouring.

**Lemma 9.2.** *Provided that the lists associated with the vertices do not run out of colours during the execution of Algorithm 9.1, then the algorithm produces a um-colouring from $\mathcal{L}$.*

*Proof.* Consider any hyperedge $S \in \mathcal{E}$. Let $t$ be the last iteration of the while loop during which some vertex of $S$ was coloured. Let $c$ denote the colour

*Strictly speaking, it is sufficient to require that $U$ is an independent subset of $V^c$ in $H[V]$. This does not play an important role, though.

chosen in that iteration. Note that $c$ is a maximal colour in $S$. We need to prove that it is also unique in $S$. Let $V^c$ denote the subset of uncoloured vertices at the start of iteration $t$ containing $c$ in their lists and let $U \subseteq V^c$ denote the independent set in $H[V^c]$ chosen to be coloured with $c$. Note that $S \cap V^c$ is a hyperedge in $H[V^c]$. We need to show that $|S \cap V^c| = 1$. Indeed, assume to the contrary that $|S \cap V^c| \geq 2$. Then, since $S \cap V^c$ is a hyperedge in $H[V^c]$ and $U$ is independent in $H[V^c]$, we must have $S \cap V^c \not\subseteq U$. Therefore, there must be a vertex $v \in (S \cap V^c) \setminus U$. This means that such a vertex $v \in S \cap V^c$ is not coloured in iteration $t$. Hence, it is coloured in a later iteration, a contradiction. □

The key ingredient, which will determine the necessary size of the lists of $\mathcal{L}$, is the particular choice of the independent set in the above algorithm. We assume that the hypergraph $H = (V, \mathcal{E})$ is hereditarily $k$-colourable (recall Definition 8.5 on page 108) for some fixed positive integer $k$. This is the case in many geometric hypergraphs. For example, hypergraphs induced by planar discs, pseudo-discs or, more generally, hypergraphs induced by regions having linear union complexity have such a hereditary colourability property for some small constant $k$ (see Section 9.2 for details). We must also put some condition on sizes of lists in the family $\mathcal{L} = \{L_v\}_{v \in V}$. With some hindsight, we require

$$\sum_{v \in V} \lambda^{|L_v|} < 1,$$

where $\lambda := \frac{k-1}{k}$. We are ready to state the main theorem.

**Theorem 9.3.** *Let $H = (V, \mathcal{E})$ be a hereditarily $k$-colourable hypergraph and set $\lambda := \frac{k-1}{k}$. Let $\mathcal{L} = \{L_v\}_{v \in V}$, such that $\sum_{v \in V} \lambda^{|L_v|} < 1$. Then, $H$ admits a unique-maximum colouring from $\mathcal{L}$.*

*Proof.* We refine Algorithm 9.1, by showing how to choose a good independent set.

A natural choice of a good independent set would of course be the largest one. Unfortunately, such a naïve approach does not work, because it does not take into account the remaining colours in the lists of the uncoloured vertices. Instead, we use a method of choosing the independent set that gives priority to colouring vertices with fewer remaining colours in their lists. Towards that goal, we define a potential function on subsets of uncoloured vertices and we choose the independent set with the highest potential (the potential quantifies how dangerous it is that some vertex in the set will run out of colours in its list).

For an uncoloured vertex $v \in V$, let $r_t(v)$ denote the number of colours remaining in the list of $v$ at the beginning of iteration $t$ of the algorithm and for coloured vertices $v \in V$, set $r_t(v) := \infty$. Obviously, the value of $r_t(v)$ depends on the particular run of the algorithm. For a subset of uncoloured vertices $X \subseteq V$ at the beginning of iteration $t$, let $P_t(X) := \sum_{v \in X} \lambda^{r_t(v)}$. We define the potential at the beginning of iteration $t$ to be $P_t := P_t(V_t)$, where $V_t$ denotes the subset of all uncoloured vertices at the beginning of iteration $t$. Note that the value of the potential at the beginning of the algorithm (i.e. in the first iteration) is $P_1 = \sum_{v \in V} \lambda^{|L_v|} < 1$.

Our goal is to show that, with the right choice of the independent set in each iteration, we can ensure that for any iteration $t$ and every vertex $v \in V_t$ the inequality $r_t(v) > 0$ holds. In order to achieve this, we will show that, with the right choice of the subset of vertices coloured in each iteration, the potential function $P_t$ is non-increasing in $t$. This will imply that for any iteration $t$ and every uncoloured vertex $v \in V_t$ we have:

$$\lambda^{r_t(v)} \leq P_t \leq P_1 < 1,$$

and hence $r_t(v) > 0$, as required.

Assume that the potential function is non-increasing until iteration $t$. Let $P_t$ be the value of the potential function in the beginning of iteration $t$ and let $c$ be the colour associated with iteration $t$. Recall that $V_t$ denotes the set of uncoloured vertices that are considered in iteration $t$, and $V^c \subseteq V_t$ denotes the subset of uncoloured vertices that contain the colour $c$ in their lists. Put $P' = P_t(V_t \setminus V^c)$ and $P'' = P_t(V^c)$. Notice that $P_t = P' + P''$. Let us describe how we find the independent set of vertices to be coloured in iteration $t$. First, we find an auxiliary proper colouring of the hypergraph $H[V^c]$ with $k$ colours. Consider the colour class $U$ which has the largest potential $P_t(U)$. Since the vertices in $V^c$ are partitioned into at most $k$ independent subsets $U_1, \ldots, U_k$ and $P'' = \sum_{i=1}^{k} P_t(U_i)$, then by the pigeon-hole principle there is an index $j$ for which $P_t(U_j) \geq P''/k$. We choose $U = U_j$ as the independent set to be coloured in iteration $t$. Notice that, in this case, the value $r_{t+1}(v) = r_t(v) - 1$ for every vertex $v \in V^c \setminus U$, and all the vertices in $U$ are coloured. For vertices in $V_t \setminus V^c$, there is no change in the size of their lists. Thus, the value $P_{t+1}$ of the potential function at the end of iteration $t$ (and in the beginning of iteration $t+1$) is $P_{t+1} \leq P' + \lambda^{-1}(1 - \frac{1}{k})P''$. Since $\lambda = \frac{k-1}{k}$, we have $P_{t+1} \leq P' + P'' = P_t$, as required. □

It is evident that Theorem 9.3 has an algorithmic version:

**Corollary 9.4.** *Let $H = (V, \mathcal{E})$ be a hypergraph on $n$ vertices which is hereditarily properly $k$-colourable and let $\mathcal{L} = \{L_v\}_{v \in V}$, such that $\sum_{v \in V} \lambda^{|L_v|} < 1$. As-*

**Algorithm 9.2** UMColour($H$, $\mathcal{L}$): Unique-max colour the hypergraph $H = (V, \mathcal{E})$ from $\mathcal{L}$

**Require:** $H$: a hereditarily properly $k$-colourable hypergraph
  $\lambda := \frac{k-1}{k}$
  **for** $v \in V$ **do**
    $r(v) \leftarrow |L_v|$
  **end for**
  **while** $V \neq \emptyset$ **do**
    $c \leftarrow \min \bigcup_{v \in V} L_v$ (* *c is the minimum colour among all lists* *)
    $V^c \leftarrow \{v \in V \mid c \in \mathcal{L}_v\}$
    compute a proper colouring of $H[V^c]$ with at most $k$ colours and colour classes $U_1, \ldots, U_k$
    $U \leftarrow$ a colour class among $U_1, \ldots, U_k$ maximising $\sum_{v \in U_i} \lambda^{r(v)}$
    **for** $x \in U$ **do**
      $f(x) \leftarrow c$
    **end for**
    **for** $v \in V^c \setminus U$ **do**
      $L_v \leftarrow L_v \setminus \{c\}$ (* *remove the colour c from all lists of uncoloured vertices in* $V^c$ *)
      $r(v) \leftarrow r(v) - 1$ (* *update the number of remaining colours from the list of* $v$ *)
    **end for**
    $V \leftarrow V \setminus U$ (* *remove the coloured vertices* *)
  **end while**
  **return** $f$

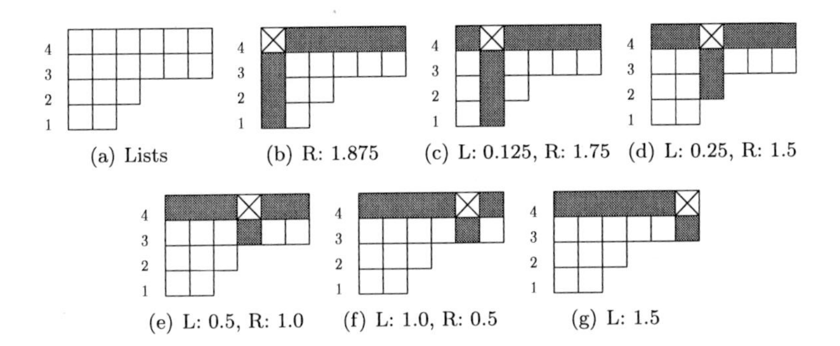

(a) Lists (b) R: 1.875 (c) L: 0.125, R: 1.75 (d) L: 0.25, R: 1.5

(e) L: 0.5, R: 1.0 (f) L: 1.0, R: 0.5 (g) L: 1.5

Figure 9.1: Lists: example (colours horizontal, lists vertical)

*sume that we have an efficient algorithm for $k$-proper-colouring every induced subhypergraph of $H$. Then, we also have an efficient algorithm for unique-maximum colouring $H$ from $\mathcal{L}$.*

**Remark 9.5.** *Some readers might be wondering, whether the potential method is a result of some derandomisation. Although it was not an intention, it turns out that there is a matching probabilistic proof, which can be derandomised by the method of conditional probabilities into Algorithm 9.2.*

*In short, one can always choose the independent set by choosing a random colour class of the auxiliary colouring instead of the colour class with the highest potential. The condition $\sum_{v \in V} \lambda^{|L_v|} < 1$ then translates into an upper bound on the probability that the algorithm runs out of colours at some vertex. Since $\lambda^{|L_v|}$ is the probability that the algorithm runs out of colours at vertex $v$, the sum is the union bound on the probability that some vertex runs out of colours. If the sum is smaller than one, the existence of a um-colouring is guaranteed.*

We also show that the conditions of Theorem 9.3 are, in a sense, best possible. Remember the discrete interval hypergraph $H_n$ examined in the beginning of this chapter. It is easy to see that $H_n$ is hereditarily 2-proper-colourable, i.e. $k = 2$ and $\lambda = 1/2$ in the notation of Theorem 9.3.

**Theorem 9.6.** *Given $n$ positive integers $a_1, a_2, \ldots, a_n$, such that $a_1 \geq a_2 \geq \cdots \geq a_n$ and $\sum_{i \in [n]} 2^{-a_i} \geq 1$, there exists a family $\mathcal{L} = \{L_i\}_{i \in [n]}$ (of sets of colours) such that $|L_i| = a_i$ for every $i \in [n]$ and the discrete interval hypergraph $H_n$ does not admit a unique-maximum colouring from $\mathcal{L}$.*

Consider an example of $a = (4, 4, 3, 2, 2, 2)$ and lists of consecutive colours, each with maximum colour 4 (see Figure 9.1). One can observe that after colouring some vertex with the highest colour and removing this colour, either the points to the left or to the right of the coloured vertex have again sum at least 1. The idea of the proof will be similar.

*Proof.* We will prove by induction on $n$ that for the family $\mathcal{L} = \{L_i\}_{i\in[n]}$ with

$$L_i = \{c \mid a_1 + 1 - a_i \leq c \leq a_1\},$$

for every $i \in [n]$, the hypergraph $H_n$ does not admit a um-colouring from $\mathcal{L}$. For $n = 1$, the above statement trivially holds.

For $n > 1$, assume for the sake of contradiction that $H_n$ admits a um-colouring $C$ from $\mathcal{L}$. Without loss of generality, the maximum colour in $C$ is $a_1$ and it occurs in exactly one vertex of $H_n$, say $m$, where $m \in [n]$.

Consider the following two discrete interval subhypergraphs of $H_n$:

$$H' \text{ induced by } [1, m-1] \quad \text{and} \quad H'' \text{ induced by } [m+1, n],$$

and the families

$$\begin{aligned} \mathcal{L}' &= \{L'_i\}_{i\in[1,m-1]} = \{L_i \setminus \{a_1\}\}_{i\in[1,m-1]} \quad \text{and} \\ \mathcal{L}'' &= \{L''_i\}_{i\in[m+1,n]} = \{L_i \setminus \{a_1\}\}_{i\in[m+1,n]}. \end{aligned}$$

Restricting colouring $C$ to each one of $H'$, $H''$ shows that $H'$ admits a um-colouring from $\mathcal{L}'$ and $H''$ admits a um-colouring from $\mathcal{L}''$.

By applying the inductive hypothesis for the smaller hypergraphs $H'$ and $H''$, we get

$$\sum_{1\leq i<m} 2^{-|L'_i|} < 1 \quad \text{and} \quad \sum_{m<i\leq n} 2^{-|L''_i|} < 1.$$

We write

$$\sum_{1\leq i\leq n} 2^{-|L_i|} = P' + 2^{-|L_m|} + P'', \tag{9.1}$$

where $P' = \sum_{1\leq i<m} 2^{-|L_i|}$ and $P'' = \sum_{m<i\leq n} 2^{-|L_i|}$. If $P'' \geq 1/2$, then

$$\sum_{m<i\leq n} 2^{-|L''_i|} = \sum_{m<i\leq n} 2^{-(|L_i|-1)} = 2P'' \geq 1,$$

which is a contradiction. Hence, $P'' < 1/2$. Moreover, $P'' \leq 2^{-1} - 2^{-|L_m|}$, because $P''$ is a sum of multiples of $2^{-|L_m|}$ (remember that $|L_m| \geq |L_i|$ for $i \in [m+1, n]$). Finally, from $\sum_{1\leq i\leq n} 2^{-|L_i|} \geq 1$ and (9.1), we get

$$P' \geq 1 - P'' - 2^{-|L_m|} \geq 1 - 2^{-1} + 2^{-|L_m|} - 2^{-|L_m|} = 1/2,$$

which implies

$$\sum_{1\leq i<m} 2^{-|L'_i|} = \sum_{1\leq i<m} 2^{-(|L_i|-1)} = 2P' \geq 1,$$

a contradiction. □

## 9.2 Geometric hypergraphs

Consider a hypergraph $H = (V, \mathcal{E})$ with $n$ vertices and a family $\mathcal{L} = \{L_v\}_{v\in V}$, such that for every $v \in V$, the list of $v$ has size $|L_v| > \log_\lambda n$. Then, $\sum_{v\in V} \lambda^{-|L_v|} < 1$ and thus we have the following special case of Theorem 9.3.

**Theorem 9.7.** *Let $H = (V, \mathcal{E})$ be a hereditarily $k$-colourable hypergraph. Let $\mathcal{L} = \{L_v\}_{v\in V}$, such that $|L_v| > \log_{\frac{k}{k-1}} n$, for every $v \in V$. Then, $H$ admits a unique-maximum colouring from $\mathcal{L}$.*

As a corollary of Theorem 9.7 we obtain asymptotically optimal bounds on the um-choice number (hence, also on the cf-choice number) of many geometric hypergraphs.

**Corollary 9.8.** *Let $C$ be some absolute constant. Let $\mathcal{R}$ be a (possibly infinite) family of simple planar Jordan regions such that, for any $n$ and any subset $\mathcal{R}' \subseteq \mathcal{R}$ of $n$ regions, the union complexity of $\mathcal{R}'$ is bounded by $Cn$. Let $H = H(\mathcal{R}')$ be a hypergraph induced by a subset $\mathcal{R}' \subseteq \mathcal{R}$ of $n$ such regions. Then*

$$ch_{\mathrm{um}}(H) = O(\log n).$$

This follows from the fact that such a hypergraph has chromatic number $O(1)$ (by Theorem 8.29) and from Theorem 9.7.

**Corollary 9.9.** *Let $\gamma > 0$ and $s \in \mathbb{N}$ be absolute constants. Let $\mathcal{R}$ be an $s$-intersecting family of $n$ compact locally $\gamma$-fat (see Definition A.6) planar regions. Then,*

$$ch_{\mathrm{um}}(H(\mathcal{R})) \leq \log n \cdot 2^{O\left(\alpha(n)^{(s-2)/2}\right)}$$

This is a consequence of Theorem 9.7 together with Theorem 8.29. The union complexity of such regions is bounded in the appendix by Theorem A.7 on page 149 and Theorem A.8 on page 149.

**Corollary 9.10.** *Let $P$ be a set of $n$ points in $\mathbb{R}$. Then*

$$ch_{\mathrm{um}}(H_{\mathcal{I}}(P)) \leq \log_2 n + 1.$$

The above hypergraph is isomorphic to the discrete interval hypergraph $H_n$ which is hereditarily 2-colourable by Proposition 8.17.

**Corollary 9.11.** *Let $P$ be a set of $n$ points in the plane. Then,*

$$ch_{\mathrm{um}}(H_{\mathcal{D}}(P)) \leq \log_{4/3} n + 1 \approx 2.41 \log_2 n + 1.$$

**Corollary 9.12.** *Let $\mathcal{D}' \subseteq \mathcal{D}$ be a set of $n$ discs. Then,*

$$ch_{\mathrm{um}}(H(\mathcal{D}')) \leq \log_{4/3} n + 1.$$

This follows from combining the fact that such hypergraphs are hereditary 4-colourable (Proposition 8.19 on page 116 and Proposition 8.20 on page 116) with Theorem 9.7.

**Corollary 9.13.** *Let $\mathcal{P}$ be a family of $n$ pseudo-discs. Then*

$$ch_{\mathrm{um}}(H(\mathcal{P})) \in O(\log n).$$

This is implied by the fact that families of pseudo-discs have chromatic number $O(1)$ as shown in Theorem 8.30 on page 118 or from their linear union complexity and Corollary 9.8.

**Corollary 9.14.** *Let $P$ be a set of $n$ points in $\mathbb{R}^2$. Then,*

$$ch_{\mathrm{um}}(H_{\mathcal{H}}(P)) \leq \log_{3/2} n + 1 \approx 1.71 \log_2 n + 1.$$

This is a consequence of the hypergraph being hereditarily 3-colorable, except when there are 4 points of which only 3 are at the convex hull (by Proposition 8.18 on page 116).

**Corollary 9.15.** *Let $P$ be a set of $n$ points in $\mathbb{R}^2$. Then,*

$$ch_{\mathrm{um}}(H_{\mathcal{A}}(P)) \in O(n^{0.382}).$$

This follows from the bound from Theorem 8.27 on page 117 (or, more precisely, its exact version from [AEGR07]).

# 10 Little pieces

This chapter is a collection of several smaller results related to um- and cf-choosability.

In Section 10.1, we give upper bounds on the online list um- and cf-colouring for intervals: $O(\log n)$ colours in the dynamic offline model and $O(\sqrt{n})$ and $\Theta(n)$ in the relative position online model for cf- and um-colouring, respectively. This reveals a large difference between the list cf- and list um-colouring which is not present without the lists.

In Section 10.2, we obtain an asymptotically tight upper bound on the cf-choice number of hypergraphs consisting of the vertices of a planar graph together with all subsets of vertices that form a simple path in the graph (see Cheilaris and Tóth [CT10] for applications of this class of hypergraphs); it is not possible to prove a similar upper bound on the um-choice number and indeed we show that the um-choice number of a hypergraph induced by paths of a planar graph can be substantially higher.

In Section 10.3, using the list colouring approach, we prove tight upper bounds on the um-choice number in terms of (a) the number of hyperedges in the hypergraph or (b) the maximum degree of a vertex. These results extend results of Cheilaris [Che09] and Pach and Tardos [PT09]. Moreover, the list colouring approach allows us to give a more concise proof.

In Section 10.4, we present a general bound on the cf-choice number of any hypergraph in terms of its cf-chromatic number. We show that for any hypergraph $H$ (not necessarily of a geometric nature) with $n$ vertices we have: $ch_{\mathrm{cf}}(H) \leq \chi_{\mathrm{cf}}(H) \cdot \ln n + 1$. The proof of this fact uses an idea which is an extension of the probabilistic proof first given by Erdős et al. [ERT80]. There, it was proved that the choice number of every bipartite graph with $n$ vertices is $O(\log n)$. The argument can be generalised to a large natural class of colourings (i.e. not just conflict-free); however, we note that such a bound is not possible for $ch_{\mathrm{um}}$.

Finally, in Section 10.5, we study the (proper) choice number of several geometric hypergraphs and show that many of the known bounds for the proper colouring of the underlying hypergraphs hold in the context of list-colouring as well.

This is joint work with Panos Cheilaris and Shakhar Smorodinsky [CSS10].

## 10.1 Online colouring points with respect to intervals

Let us briefly recall the models of dynamic colouring hypergraphs introduced in Section 8.3.4. We describe their list colouring counterparts. A sequence $(H^t)_{t=0}^n$ of hypergraphs is given together with lists $\mathcal{L} = \{L_v\}_{v \in V(H^n)}$, where $H^t$ has $t$ vertices and additionally, $H^{t-1}$ is an induced subhypergraph of $H^t$. A conflict-free colouring $C^t : V(H^t) \to \mathbb{N}$ of $H^t$ from the lists $\mathcal{L}^t := \{L_v\}_{v \in V(H^t)}$ extending the previous colouring $C^{t-1}$ (i.e. $C^t|_{V(H^{t-1})} = C^{t-1}$) has to be found.

In the *dynamic offline model*, the whole sequence $(H^t)$ as well as the lists $\mathcal{L}$ are known in advance. In the online models, the hypergraphs and their lists are revealed at discrete timesteps and the new vertex $v^t$ has to be coloured without knowing the future hypergraphs. In the *dynamic online-with-absolute-positions model* the final hypergraph with the lists is known in advance and the vertices are labelled. In the *dynamic online-with-relative-positions model* nothing, except possibly the structure of the hypergraphs, is known (e.g. we know that we are colouring points in $\mathbb{R}$ with respect to intervals). We are interested in the minimum size of lists that suffice for finding such a colouring.

Our framework in Algorithm 9.2 and the hereditary 3-colourability of the underlying hypergraph (Lemma 8.32) imply the following:

**Corollary 10.1.** *Lists of size* $\log_{3/2} n + 1$ *are sufficient for dynamic offline unique-maximum colouring of points on the real line with respect to intervals.*

This is the same result as we know in the non-list situation ([BNCS06,

BNCS08]). In the rest of this subsection, we observe that the situation changes dramatically, when we look at the relative positions online model: even if the adversary is forced to reveal the (unknown) points in an increasing order, i.e. $v^{t-1} < v^t$. That means that the hypergraph to be coloured is simply the discrete interval hypergraph on some (known or unknown) number of vertices but the lists are not known in advance. We show the following:

**Theorem 10.2.** *Online list um-colouring of $n$ points on a real line in the relative-positions model might require lists of size $n$ even if the points have to be revealed in an increasing order.*

*Proof.* Assume that $n-1$ colours suffice. The following simple strategy for the adversary prevents the last point from getting coloured: Let $x_1 < x_2 < \ldots < x_n$ be the points in an increasing order. Define the lists $L_{x_i} := \{i \cdot n + 1, \ldots, (i+1) \cdot n - 1\}^*$ for $i \in [n-1]$. The colouring $C$ has to colour $x_1, \ldots x_{n-1}$ by a increasing sequence of $n-1$ distinct colours. The last point $x_n$ needs to be coloured by a different colour: if $C(x_i) = C(x_n)$, then the interval $[x_i, x_n]$ does not have a unique maximum. Consequently, if the adversary gives a list $L_{x_n} := \{C(x_1), \ldots, C(x_{n-1})\}$, the colouring cannot be extended to $x_n$. □

This shows that list um-colouring is hopeless in the relative positions model even for intervals and when we constrain the adversary. Fortunately, this is not the case for the conflict-free colouring. Let us first show that for the constrained adversary (who presents the points in an increasing sequence), we can do much better.

**Observation 10.3.** *Lists of size $\lceil\sqrt{2n+1}\rceil - 1$ are sufficient for online list cf-colouring of $n$ points on a real line in the relative-positions model with the restriction that the points are revealed in an increasing order.*

*Proof.* Let $k$ denote the size of the lists. We propose the following colouring strategy: Split the input into $k$ rounds $R_1, \ldots, R_k$, where $R_i$ consists of $k+1-i$ points. Within each round we use distinct colours. The colour of the last point in $C_{R_i}$ of every round $i$ is declared unique for the rounds $R_i, R_{i+1}, \ldots$ and is forbidden in the following rounds. Until the round $R_{i+1}$, we have forbidden $i$ colours but we insist on colouring only $k-i$ points with distinct colours in $R_i$ which can be done from any lists.

We show that this strategy indeed produces a cf-colouring. Consider a non-empty interval: if it only contains points from a single round, then all the colours are different and hence, the interval has a unique colour. Otherwise,

*The only important property is that they are disjoint and all colours in $L_{x_a}$ are smaller than all colours in $L_{x_b}$ for $a \leq b$.

it contains points of several rounds $R_i, R_{i+1}, \ldots, R_j$. Then the colour $C_{R_i}$ must be unique in this interval: by the algorithm, it is unique in $R_i$ and was forbidden in all subsequent rounds. Thus, the resulting colouring is conflict-free. Such an algorithm can colour up to $\binom{k+1}{2}$ points from lists of size $k$ and hence, we need $\lceil\sqrt{2n+1}\rceil - 1$ colours for $n$ points. □

The main idea of the algorithm was introducing a separating unique colour after approximately every $\sqrt{n}$ points. A closer look reveals that a similar idea also works when we drop the restriction on the adversary (namely that the points are revealed in an increasing order). Below, we give an informal description of the approach which is then summarised in Algorithm 10.1.

**Algorithm 10.1** CFOnlineInterval: Online conflict-free list colour points with respect to intervals

```
F ← {(−∞, ∞)}  (* Interval subdivision of the ℝ *)
F((−∞, ∞)) ← ∅  (* Forbidden colours for the interval *)
P((−∞, ∞)) ← all points  (* Points lying in the interval *)
for t = 1, ... do
  x ← the point revealed at time t
  L_x ← the list of x
  (a, b) ← the interval of F where x lies
  if |L_x| − |F((a, b))| > |P((a, b))| then
    (* Colour that is neither forbidden nor used in (a, b) *)
    C(v) ← any colour from L_x \ (C(P((a, b))) ∪ F((a, b)))
  else
    m ← median of P((a, b))

    (* Split the interval and forbid C(m) in both subintervals *)
    F ← F \ {(a, b)} ∪ {(a, m), (m, b)}
    F((a, m)) ← F((a, b)) ∪ {C(m)}
    P((a, m)) ← {u ∈ P((a, b)): u < m}
    F((m, b)) ← F((a, b)) ∪ {C(m)}
    P((m, b)) ← {u ∈ P((a, b)): u > m}
    (a', b') ← the interval of F where x lies

    (* C(x) is any colour — neither forbidden for nor used in (a', b') *)
    C(x) ← any colour from L_x \ (C(P((a', b'))) ∪ F((a', b')))
  end if
end for
```

If the number of points we coloured is smaller than the size $k = |L_x|$ of the list, we can colour the new point $x$ by a new colour. Otherwise, pick the median $m$ of the coloured points and declare its colour to be unique (i.e. remove the colour $C(m)$ from all the lists we will see from now on, including $L_x$). This splits the whole problem into two independent subproblems: (i) points to the left of $m$, and (ii) points to the right of $m$. The point $x$ falls into one of these subproblems. We have an available colour for colouring $x$ uniquely in that subproblem unless there were only 2 or fewer points (which means that we only had $|L_x| \leq 2$ and hence, we must have forbidden almost all colours already).

Generally speaking, whenever we see a new point, it falls into one of the subproblems and the forbidden colours are subproblem-specific (i.e. forbidding a colour $c$ for all future points when we are in the subproblem $(a, b)$ only means that the points inside $(a, b)$ cannot use $c$ but all the points outside $(a, b)$ can).

Assume we have lists of size $k$ at our disposal. If Algorithm 10.1 splits an interval $(a, b)$ at some point $m$ at time $t$, then it does not make sense for the adversary to present points in both subintervals $(a, m)$ and $(m, b)$ any more. Every interval intersecting both $(a, m)$ and $(m, b)$ has a uniquely coloured point: either $m$ or some earlier splitting point. Hence, the first hyperedge that is not cf-coloured has to lie in one of the subproblems $(a, m)$ and $(m, b)$. Without loss of generality, it lies in $(a, m)$. But then ignoring all the points in $(m, b)$ after the time $t$ also results in the same non-cf-coloured hyperedge, after presenting fewer points.

We can conclude that every time we split an interval into two subintervals, the adversary decides on one of the subintervals and all the following points will lie in that subinterval.

Now, we are in a very similar situation as in Observation 10.3. We have rounds $R_1, R_2, \ldots$ (between the splits) in which we use distinct colours. In the first round, $k$ points are revealed. In each subsequent round $R_i$, we already have up to $\lceil \frac{k-i+1}{2} \rceil$ points pre-coloured before the split and hence, we can only colour another $\lfloor \frac{k-i+1}{2} \rfloor$ points before we run out of colours for that round. The algorithm can have up to $k - 1$ rounds before it fails to continue. Thus, we colour a total of $\lfloor \frac{k}{2} \rfloor + \sum_{i=0}^{k-2} \lfloor \frac{k-i}{2} \rfloor$ points which is $2\binom{\frac{k+1}{2}}{2} + \frac{k-1}{2}$ for $k$ odd and $2\binom{\frac{k}{2}}{2} + k - 1$ for $k$ even. Both of the expressions are lower bounded by $\frac{k^2}{4}$ for $k \geq 2$. This implies the following:

**Theorem 10.4.** *Lists of size $2\lceil \sqrt{n} \rceil$ are sufficient for online list cf-colouring of $n$ points on a real line in the relative-positions model, if $n \geq 3$.*

This reveals a striking difference between the um- and cf-choice numbers which is not present for the usual chromatic numbers.

**Remark 10.5.** *We do not know any better lower bound than $\Omega(\log n)$ for either of the previous settings.*

## 10.2 List cf-colouring of planar graphs with respect to paths

Recall the definition of a cf-colouring with respect to paths. Given a simple graph $G = (V, E)$, consider the hypergraph

$$H_G^{\text{path}} = (V, \{S \mid S \text{ is the vertex set of a simple path in } G\}).$$

**Theorem 10.6.** *Let $G$ be a planar graph with $n$ vertices. Then its conflict-free path choice number is $ch_{\text{cf}}(H_G^{\text{path}}) \in O(\sqrt{n})$.*

*Proof.* The proof is constructive. Given a planar graph $G$ on $n$ vertices together with a family $\mathcal{L} = \{L_v\}_{v \in V}$ of sets of size $c\sqrt{n}$ where $c$ is some absolute constant to be revealed later, we produce a cf-colouring $C$ of $G$ with respect to paths with colours from $\mathcal{L}$.

The algorithm is recursive. By the Lipton-Tarjan separator theorem [LT80] and in particular by the version of the separator theorem of Djidjev [Dji82], there exists a partition of the vertex set $V = R \cup B \cup S$ such that $\max(|R|, |B|) \leq 2n/3$ and $|S| \leq \sqrt{6n}$ and such that there is no edge connecting a vertex in $R$ with a vertex in $B$. Moreover, this partition can be computed efficiently.

We colour all vertices in $S$ by distinct colours. This can be done greedily as follows: Arbitrarily order the vertices in $S$ and for each vertex $v$ in this order choose a colour from $L_v$ to be assigned to $v$ which is distinct from all the colours assigned to the previous vertices in $S$. This is possible if $|L_v| = c\sqrt{n} \geq \sqrt{6n} \geq |S|$. Next, for each vertex $u \in R \cup B$ modify the lists $\{L_u\}_{u \in R \cup B}$ by erasing all the colours used for $S$, namely put $\mathcal{L}' = \{L_u \setminus \{C(v) \mid v \in S\}\}_{u \in R \cup B}$. We recursively colour $G[B]$ and $G[R]$ from $\mathcal{L}'$. Note that the colours assigned to the vertices in $R \cup B$ are distinct from all the colours used for $S$. Note also that if this colouring is indeed a valid cf-colouring of $G$ from $\mathcal{L}$ then the function $f(n)$ defined to be the maximum cf-choice number of a planar graph on $n$ vertices satisfies the following recursive inequality:

$$\begin{aligned} f(n) &\leq \sqrt{6}\sqrt{n} + f(2n/3) \leq \sum_{i=0}^{\infty} \sqrt{6 \cdot \left(\tfrac{2}{3}\right)^i n} \\ &= \frac{\sqrt{6}\sqrt{n}}{1-\sqrt{2/3}} \approx 13.3485\sqrt{n}. \end{aligned}$$

Thus, we have $f(n) \leq c\sqrt{n}$, for $c \approx 13.3485$, as claimed. □

**Remark 10.7.** *The upper bound $O(\sqrt{n})$ is asymptotically tight, since for the $\sqrt{n} \times \sqrt{n}$ grid graph $G_{\sqrt{n}}$, it was proved by Cheilaris and Tóth [CT10] that $\chi_{\mathrm{cf}}(H^{\mathrm{path}}_{G_{\sqrt{n}}}) = \Omega(\sqrt{n})$ and thus, from inequality (8.3), also $ch_{\mathrm{cf}}(H^{\mathrm{path}}_{G_{\sqrt{n}}}) = \Omega(\sqrt{n})$.*

It is easily seen that an analogue of Theorem 10.6 for $ch_{\mathrm{um}}$ does not hold. For example, consider the *star graph* on $n > 2$ vertices $K_{1,n-1}$. It is easy to check that $\chi_{\mathrm{cf}}(H^{\mathrm{path}}_{K_{1,n-1}}) = ch_{\mathrm{cf}}(H^{\mathrm{path}}_{K_{1,n-1}}) = \chi_{\mathrm{um}}(H^{\mathrm{path}}_{K_{1,n-1}}) = 2$. However, consider a family $\mathcal{L}$ of lists as follows: Associate the $n-1$ leaves of the star with the same list of colours and associate the centre vertex with a list of colours which are all lower than the colours appearing in the lists of the leaves. In a unique-maximum colouring with respect to paths from $\mathcal{L}$, no two leaves can get the same colour, because then we do not have the unique maximum property for the path that connects these two leaves and hence, $ch_{\mathrm{um}}(H^{\mathrm{path}}_{K_{1,n-1}}) \geq n-1$.

## 10.3 List um-colouring hypergraphs with few edges

In this section we extend upper bounds on $\chi_{\mathrm{cf}}$ of Cheilaris [Che09] and Pach and Tardos [PT09], making them hold also for $ch_{\mathrm{um}}$. In other words, we extend the results in two ways making them hold for choice instead of chromatic number and for unique-maximum colourings instead of conflict-free colourings. Moreover, we give a more concise proof. In order to state the results, we need the following definition.

**Definition 10.8.** *For every hypergraph $H$, define $s(H)$ to be the minimum positive integer $s$ such that $|E(H)| \leq \binom{s}{2}$.*

**Fact 10.9.** *For two hypergraphs $H$ and $H'$, if $|\mathcal{E}(H)| \geq |\mathcal{E}(H')|$ and $s(H) = s(H') > 1$, then*

$$|\mathcal{E}(H)| - |\mathcal{E}(H')| < s(H) - 1.$$

*Proof.* We have $\binom{s(H)-1}{2} < |\mathcal{E}(H')| \leq |\mathcal{E}(H)| \leq \binom{s(H)}{2}$ which in turn implies $|\mathcal{E}(H)| - |\mathcal{E}(H')| < s(H) - 1$. □

**Theorem 10.10.** *Let $H = (V, \mathcal{E})$ be a hypergraph and let $\mathcal{L} = \{L_v\}_{v \in V}$ be a family of lists. If, for every $v \in V$, $|L_v| \geq \min(\deg_H(v) + 1, s(H))$, then $H$ admits a unique-maximum colouring from $\mathcal{L}$.*

*Proof.* Notice that, if $s(H) = 1$, the hypergraph has no hyperedge, and thus, if $|L_v| \geq 1$, for every $v \in V$, then $H$ admits a unique-maximum colouring from

$\mathcal{L}$. The proof is by induction on $|V|$. If $H$ has one vertex $v$, then $\deg_H(v) = 0$ and $s(H) = 1$. Hence, if $|L_v| \geq 1$, then $H$ admits a unique-maximum colouring from $\mathcal{L}$.

If $|V| > 1$ and $s(H) > 1$, consider the maximum colour occurring in the union of all lists, that is, $c = \max \bigcup_{v \in V} L_v$. Among those vertices which have $c$ in their list, choose the vertex $v$ with maximum degree in the hypergraph. Consider the subset of hyperedges $\mathcal{E}_v \subseteq \mathcal{E}$ that contain $v$. Put $H' = (V', \mathcal{E}')$, where $V' = V \setminus \{v\}$, $\mathcal{E}' = \mathcal{E} \setminus \mathcal{E}_v$, and define $\mathcal{L}' = \{L'_u\}_{u \in V'}$ such that

$$L'_u = \begin{cases} L_u \setminus \{c\} & \text{if } u \in \bigcup_{S \in \mathcal{E}_v} S, \\ L_u & \text{if } u \notin \bigcup_{S \in \mathcal{E}_v} S. \end{cases}$$

In order to apply the inductive hypothesis on $H'$, we prove that for every $u \in V'$,

$$|L'_u| \geq \min(\deg_{H'}(u) + 1, s(H')), \tag{10.1}$$

If $|L_u| = |L'_u|$, that is, when $u \notin \bigcup_{S \in \mathcal{E}_v} S$ or $c \notin L_u$, then condition (10.1) holds. Also, if $c \in L_u$, $u \in \bigcup_{S \in \mathcal{E}_v} S$, and $s(H') < s(H)$, then condition (10.1) holds, since $\deg_{H'}(u) < \deg_H(u)$. If $c \in L_u$, $u \in \bigcup_{S \in \mathcal{E}_v} S$, and $s(H') = s(H)$, then Fact 10.9 implies

$$s(H) > |\mathcal{E}| - |\mathcal{E}'| + 1 = \deg_H(v) + 1 \geq \deg_H(u) + 1.$$

This follows from the fact that $\deg_H(v) = |\mathcal{E}| - |\mathcal{E}'|$ and $\deg_H(v) \geq \deg_H(u)$. Since $s(H') = s(H)$ and $\deg_H(u) \geq \deg_{H'}(u) + 1$, we also have

$$s(H') > \deg_{H'}(u) + 1.$$

As a result,

$$\begin{aligned} |L'_u| = |L_u| - 1 \geq \min(\deg_H(u) + 1, s(H)) - 1 = \deg_H(u) + 1 - 1 \\ \geq \deg_{H'}(u) + 1 = \min(\deg_{H'}(u) + 1, s(H')). \end{aligned}$$

Finally, by the inductive hypothesis, $H'$ admits a um-colouring from $\mathcal{L}'$. Extend this colouring by colouring $v$ with $c$ to get a um-colouring of $H$. $\square$

**Corollary 10.11.** *For every hypergraph $H$, $ch_{\mathrm{um}}(H) \leq \Delta(H) + 1$.*

**Corollary 10.12.** *For every hypergraph $H$, $ch_{\mathrm{um}}(H) \leq s(H)$.*

## 10.4 Choosability and colourability in general hypergraphs

**Definition 10.13.** *We call $C'$ a* refinement *of a colouring $C$ if $C(x) \neq C(y)$ implies $C'(x) \neq C'(y)$. A class $\mathcal{C}$ of colourings is said to have* the refinement property *if every refinement of a colouring in the class is also in the class.*

The class of conflict-free colourings and the class of proper colourings of a fixed hypergraph are examples of classes which have the refinement property. On the other hand, the class of unique-maximum colourings does not have this property.

For a class $\mathcal{C}$ of colourings we can define as usual the notions of chromatic number $\chi_{\mathcal{C}}$ and choice number $ch_{\mathcal{C}}$. Then, we can prove the following theorem for classes with the refinement property.

**Theorem 10.14.** *For every class of colourings $\mathcal{C}$ that has the refinement property and every hypergraph $H$ with $n$ vertices, $ch_{\mathcal{C}}(H) \leq \chi_{\mathcal{C}}(H) \cdot \ln n + 1$.*

*Proof.* If $k := \chi_{\mathcal{C}}(H)$, then there exists a $\mathcal{C}$-colouring $C$ of $H$ with colours $\{1, \ldots, k\}$, which induces a partition of $V$ into $k$ classes: $V_1 \cup V_2 \cup \cdots \cup V_k$. Consider a family $\mathcal{L} = \{L_v\}_{v \in V}$, such that for every $v$, $|L_v| = k^* > k \cdot \ln n$. We wish to find a family $\mathcal{L}' = \{L'_v\}_{v \in V}$ with the following properties:

(i) For every $v \in V$, $L'_v \subseteq L_v$.
(ii) For every $v \in V$, $L'_v \neq \emptyset$.
(iii) For every $i \neq j$, if $v \in V_i$ and $u \in V_j$, then $L'_v \cap L'_u = \emptyset$.

Obviously, if such a family $\mathcal{L}'$ exists, then there exists a $\mathcal{C}$-colouring from $\mathcal{L}'$: For each $v \in V$, pick a colour $x \in L'_v$ and assign it to $v$.

We create the family $\mathcal{L}'$ randomly as follows: Assign each element in $\cup\mathcal{L}$ uniformly at random to one of the $k$ classes of the partition $V_1 \cup \cdots \cup V_k$. For every vertex $v \in V$, say with $v \in V_i$, we create $L'_v$ by keeping only elements of $L_v$ that were assigned through the above random process to $v$'s class, $V_i$.

The family $\mathcal{L}'$ obviously has properties 1 and 3. We will prove that with positive probability it also has property 2.

For a fixed $v$, the probability that $L'_v = \emptyset$ is at most

$$\left(1 - \frac{1}{k}\right)^{k^*} \leq e^{-k^*/k} < e^{-\ln n} = \frac{1}{n}$$

and therefore, using the union bound, the probability that, for at least one vertex $v$, $L'_v = \emptyset$, is at most

$$n\left(1-\frac{1}{k}\right)^{k^*} < 1.$$

Thus, there is at least one family $\mathcal{L}'$ where property 2 also holds, as claimed. □

**Corollary 10.15.** *For every hypergraph $H$, $ch_{\text{cf}}(H) \leq \chi_{\text{cf}}(H) \cdot \ln n + 1$.*

**Corollary 10.16.** *For every hypergraph $H$, $ch(H) \leq \chi(H) \cdot \ln n + 1$.*

The argument in the proof of Theorem 10.14 is a generalisation of an argument first given in [ERT80], proving that any bipartite graph with $n$ vertices is $O(\log n)$-choosable (see also [Alo92]).

We cannot have an analogue of Theorem 10.14 for unique maximum colourings. Again, as at the end of Section 10.2, the counterexample is the hypergraph with respect to paths of the star graph, $H^{\text{path}}_{K_{1,n-1}}$, for which we have $\chi_{\text{um}}(H^{\text{path}}_{K_{1,n-1}}) = 2$, whereas $ch_{\text{um}}(H^{\text{path}}_{K_{1,n-1}}) \geq n-1$.

## 10.5 Choosability of geometric hypergraphs

In this section, we provide near-optimal upper bounds on the choice number of several geometric hypergraphs. Recall the definition of the Delaunay graph of a hypergraph: For a hypergraph $H = (V, \mathcal{E})$, the Delaunay graph $G(H)$ is the graph on the vertex set $V$ whose edges are all hyperedges of $\mathcal{E}$ of cardinality two.

**Theorem 10.17.** *The following statements hold:*

(i) *Let $H$ be a hypergraph induced by a finite set of points in the plane with respect to discs. Then $ch(H) \leq 5$.*

(ii) *Let $\mathcal{D}' \subseteq \mathcal{D}$ be a finite family of discs in the plane. Then $ch(H(\mathcal{D}')) \leq 5$.*

(iii) *Let $\mathcal{R}$ be a set of $n$ regions and let $\mathcal{U} : \mathbb{N} \to \mathbb{N}$ be a function such that $\mathcal{U}(m)$ is the maximum complexity of any $k$ regions in $\mathcal{R}$ over all $k \leq m$, for $1 \leq m \leq n$. We assume that $\frac{\mathcal{U}(m)}{m}$ is a non-decreasing function. Then, $ch(H(\mathcal{R})) = O(\frac{\mathcal{U}(n)}{n})$.*

*Proof.* (i) Consider the Delaunay graph $G = G(H_{\mathcal{D}}(P))$ on $P$, where two points $p$ and $q$ form an edge in $G$ if and only if there exists a disc $D$ such that $D \cap P = \{p, q\}$. That is, there exists a disc $D$ that cuts off $p$ and $q$ from $P$. The proof of (i) follows easily from the following known facts:

(a) Every disc containing at least two points of $P$ must also contain a Delaunay edge $\{p,q\} \in E(G)$ (by shrinkability of discs; see Section 8.1.3).
(b) $G$ is planar (see, e.g. [dBCvKO08]).
(c) Every planar-graph is 5-choosable [Tho94].

(ii) The proof of the second part follows from a reduction to three dimensions from [Smo07] and Thomassen's result [Tho94].
(iii) For the third part of the theorem, we need the following lemma from [Smo07]:

**Lemma 10.18** ([Smo07]). *Let $\mathcal{R}$ be a set of $n$ regions and let $\mathcal{U} : \mathbb{N} \to \mathbb{N}$ be a function such that $\mathcal{U}(m)$ is the maximum complexity of any $k$ regions in $\mathcal{R}$ over all $k \leq m$, for $1 \leq m \leq n$. Then, the Delaunay graph $G$ of the hypergraph $H = H(\mathcal{R})$ has a vertex with degree at most $c\frac{\mathcal{U}(n)}{n}$ where $c$ is some absolute constant.*

The proof is similar to the proof of [Smo07] of the fact that $\chi(H(\mathcal{R})) = O(\frac{\mathcal{U}(n)}{n})$. Using Lemma 10.18, we prove that $ch(H(\mathcal{R})) \leq c \cdot \frac{\mathcal{U}(n)}{n} + 1$. Let $\mathcal{L} = \{L_r\}_{r \in \mathcal{R}}$ be the sets associated with the regions of $\mathcal{R}$. The proof is by induction on $n$. Let $r \in \mathcal{R}$ be a region with at most $c \cdot \frac{\mathcal{U}(n)}{n}$ neighbours in $G$. By the induction hypothesis, the hypergraph $H(\mathcal{R} \setminus \{r\})$ has choice number $c \cdot \frac{\mathcal{U}(n-1)}{n-1} + 1 \leq c \cdot \frac{\mathcal{U}(n)}{n} + 1$ (by our monotonicity assumption on $\frac{\mathcal{U}(n)}{n}$). We need to choose a colour (out of the $c \cdot \frac{\mathcal{U}(n)}{n} + 1$ colours that are available for us in the set $L_r$) for $r$ such that the colouring of $\mathcal{R}$ is valid. Obviously, the points that are not covered by $r$ are not affected by the colouring of $r$. Note also that any point $p \in r$ that is contained in at least two regions of $\mathcal{R} \setminus r$ is not affected by the colour of $r$ since, by induction, the set of regions in $\mathcal{R} \setminus \{r\}$ containing such points is non-monochromatic. Thus we only need to colour $r$ with a colour that is different from the colours of all regions $r' \in \mathcal{R} \setminus r$, for which there is a point $p$ contained only in $r \cap r'$. However, by our choice of $r$, there are at most $c \cdot \frac{\mathcal{U}(n)}{n}$ such regions. Thus, we can assign to $r$ a colour among the $c \cdot \frac{\mathcal{U}(n)}{n} + 1$ colours available to us in $L_r$ and keep the colouring of $\mathcal{R}$ proper. This completes the inductive step. □

**Corollary 10.19.** *Let $\mathcal{P}$ be a family of $n$ pseudo-discs. Then $ch(H(\mathcal{P})) = O(1)$.*

The corollary follows immediately from the fact that such a family $\mathcal{P}$ has linear union complexity [KLPS86], combined with Theorem 10.17.

# Appendix

# Further background

## A.1 Union complexity

In Chapter 8 we mentioned several types of families of planar objects with near-linear union complexity, without really defining them. We try to expiate this transgression here and attempt to convince the reader that most reasonably looking families without too skinny objects have near-linear union complexity.

We have already mentioned earlier that any family $\mathcal{P}$ of $n$ pseudo-discs (or discs) has a linear union complexity $\mathcal{U}(\mathcal{P}) \in O(n)$. The other families from Table 8.1 on page 118 need a proper definition.

The simplest objects whose union complexity is well understood and close to linear are fat triangles.

**Definition A.1** (Fat triangles). *Let $\gamma > 0$ be some positive constant. A triangle $T$ is called* $\gamma$-fat *if all its angles are larger than $\gamma$.*

Matoušek et al. [MPS+94] studied complexity of computing the union of a family $\mathcal{R}$ of $\gamma$-fat triangles and, on their way, obtained the following bounds on their complexity:

**Theorem A.2** ([MPS+94]). *Let $\gamma > 0$ be a constant. Then the union complexity of a family $\mathcal{R}$ of $n$ $\gamma$-fat triangles is at most $\mathcal{U}(\mathcal{R}) \in O(n \log\log n)$.**

*The constant of proportionality depends on $\gamma$.

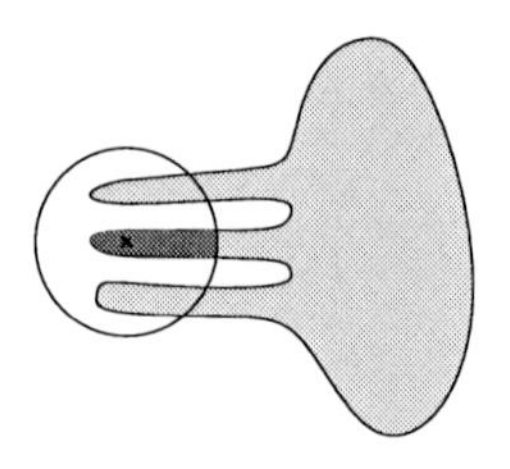

Figure A.1: The $\sqcap$ notation

*On the other hand, there are families of $\gamma$-fat triangles with union complexity $\Omega(n\alpha(n))$.*

**Theorem A.3** ([MPS+94]). *Let $\gamma > 0$ and $0 < c \leq C$ be fixed constants. Let $\mathcal{R}$ be a family of $\gamma$-fat triangles such that $c \leq \operatorname{diam}(T) \leq C$ for every triangle $T \in \mathcal{R}$. Then the boundary complexity of $\mathcal{R}$ is $O(n)$.*[1]

Until very recently, those were the best bounds. Aronov et al. [EAS11] recently obtained the following improvement:

**Theorem A.4** ([EAS11]). *Let $\gamma > 0$ be a constant. Then the union complexity of a family $\mathcal{R}$ of $n$ $\gamma$-fat triangles is at most $\mathcal{U}(\mathcal{R}) \in O(n2^{\alpha(n)} \log *n)$.*

It was a challenge to prove similar results for objects which are curved or non-convex. The first result of this type was proved by Efrat and Sharir [ES00] for fat convex objects with curved boundaries.

Later, more general concepts of families with low union complexity have been studied, including $(\alpha, \beta)$-covered objects studied by Efrat [Efr05]. Better bounds were later obtained by de Berg [dB08] for even more general class of regions, so called locally fat regions. These are currently the most general class of regions where near-linear union complexity has been proved. To define them, we will need some auxiliary notation.

**Definition A.5.** *Let $D$ be a disc with a centre $o$ and $R$ be a region in the plane. If $R$ is not convex, the intersection $R \cap D$ can consist of several connected components. The expression $D \sqcap R$ denotes the component of $D \cap R$ which contains the centre $o$ of the disc $D$ (see Figure A.1).*

This allows us to define locally $\gamma$-fat regions.

[1]The constant of proportionality depends on $\gamma, c$ and $C$.

**Definition A.6** ([dB08]). *Let $R$ be a region in the plane and $\gamma$ a parameter with $0 \leq \gamma \leq 1$. We say that $R$ is* locally $\gamma$-fat *if, for any disc $D \in \mathcal{D}$ whose centre lies in $R$ and that does not fully contain $R$ in its interior, we have* $\text{area}(D \sqcap R) \geq \gamma \cdot \text{area}(D)$.

For example, every square is locally $(2\pi)^{-1}$-fat and, more generally, a rectangle with side length ratio $\alpha$ is locally $\frac{\alpha}{\pi(1+\alpha^2)}$-fat.

The definition is very general and encompasses many kinds of shapes that are not too degenerate. The currently best published bound on the union complexity of locally fat objects was proved by de Berg [dB10]. A recent unpublished draft of Aronov yields further improvement.

**Theorem A.7** ([dB10]). *Let $\gamma > 0$ be a constant and $s \in \mathbb{N}$. Let $\mathcal{R}$ be an $s$-intersecting family of compact $\gamma$-fat regions. Then the union complexity of $\mathcal{R}$ is $\mathcal{U}(\mathcal{R}) \in O(\lambda_{s+2}(n) \log n)$.*

Here, $\lambda_s(n)$ denotes the maximum length of a Davenport-Schinzel sequence of order $s$ on $n$ symbols (for further information, see, e.g. [ASS89, DS65, HS84, HS86, Mat02]). The best lower bound construction of such family $\mathcal{R}$ yields union complexity $\mathcal{U}(\mathcal{R}) \in \Omega(\lambda_{t+2}(n))$ and this is conjectured to be the truth. The upper bounds on $\lambda$ imply:[¬]

**Theorem A.8** ([Niv10]). *Let $s \in \mathbb{N}$ be a constant. Then*

$$\lambda_s(n) \leq n \cdot 2^{O\left(\alpha(n)^{(s-2)/2}\right)}.$$

# A.2 Binomial coefficients

**Fact 1.8.** *Let $a, b \in \mathbb{Z}_0^+$. Then*

$$\sum_{i=0}^{b} \binom{a+i}{a} = \binom{a+b+1}{a+1}.$$

*Proof.* Proceed by induction on $b$. The statement holds trivially for $b = 0$ since $\binom{a}{a} = 1 = \binom{a+1}{a+1}$. For $b \geq 1$ we have $\sum_{i=0}^{b} \binom{a+i}{a} = \sum_{i=0}^{b-1} \binom{a+i}{a} + \binom{a+b}{a}$. By induction, this is $\binom{a+b}{a+1} + \binom{a+b}{a}$ which equals $\binom{a+b+1}{a+1}$ by Fact 1.7. □

[¬]The result specifies the constants inside the $O$-notation and gives a better bound for odd $s$ than we claim. We rather chose a path of a simple-to-write corollary.

**Fact 1.9.** *Let $N$ be a binomially distributed random variable $N \sim \mathrm{B}(n,p)$ and let $s \in \mathbb{N}$. Then*

$$\mathbf{E}\left[\binom{N}{s}\right] = p^s \binom{n}{s}.$$

*Proof.*

$$\begin{aligned}
\mathbf{E}\left[\binom{N}{s}\right] &= \sum_{k=s}^{n} \binom{k}{s} \cdot \binom{n}{k} p^k (1-p)^{n-k} \\
&= \sum_{k=s}^{n} \frac{k! \cdot n!}{s!(k-s)! \cdot k!(n-k)!} p^k (1-p)^{n-k} \\
&= \sum_{k=s}^{n} \frac{n! \cdot (n-s)!}{(n-s)!s! \cdot (k-s)!(n-k)!} p^k (1-p)^{n-k} \\
&= \sum_{k=s}^{n} \binom{n}{s} \binom{n-s}{k-s} p^k (1-p)^{n-k} \\
&= p^s \binom{n}{s} \sum_{k=s}^{n} \binom{n-s}{k-s} p^{k-s} (1-p)^{n-s-(k-s)}.
\end{aligned}$$

The sum is equal to 1 since we are summing the probabilities of all the possible values of a $\mathrm{B}(n-s,p)$ distributed random variable. This proves the statement.

□

# Bibliography


[AAHP⁺98] Artur Andrzejak, Boris Aronov, Sariel Har-Peled, Raimund Seidel, and Emo Welzl. Results on $k$-sets and $j$-facets via continuous motion. In *Symposium on Computational Geometry*, pages 192–199, 1998.

[ABFK92] Noga Alon, Imre Bárány, Zoltán Füredi, and Daniel J. Kleitman. Point selections and weak e-nets for convex hulls. *Combinatorics, Probability & Computing*, 1:189–200, 1992.

[ÁBFM⁺08] Bernardo M. Ábrego, József Balogh, Silvia Fernández-Merchant, Jesús Leaños, and Gelasio Salazar. An extended lower bound on the number of (k)-edges to generalized configurations of points and the pseudolinear crossing number of $\mathrm{k_n}$. *J. Comb. Theory, Ser. A*, 115(7):1257–1264, 2008.

[ABG⁺05] Manuel Abellanas, Prosenjit Bose, Jesús García, Ferran Hurtado, Mariano Nicolás, and Pedro A. Ramos. On properties of higher-order Delaunay graphs with applications. In *EuroCG*, pages 119–122. Technische Universiteit Eindhoven, 2005.

[ÁCFM⁺10] Bernardo M. Ábrego, Mario Cetina, Silvia Fernández-Merchant, Jesús Leaños, and Gelasio Salazar. 3-symmetric and 3-decomposable geometric drawings of $\mathrm{k_n}$. *Discrete Applied Mathematics*, 158(12):1240–1258, 2010.

[ÁCFM⁺11] Bernardo M. Ábrego, Mario Cetina, Silvia Fernández-Merchant, Jesús Leaños, and Gelasio Salazar. On ($\leq k$)-edges, crossings, and halving lines of geometric drawings of $K_n$. *arXiv*, 1102.5065, 2011.

[ACNS82] Miklós Ajtai, Vašek Chvátal, Monroe M. Newborn, and Endre Szemerédi. Crossing-free subgraphs. In *Theory and practice of combinatorics*, volume 60 of *North-Holland Math. Stud.*, pages 9–12. North-Holland, Amsterdam, 1982.

[AEGR07] Deepak Ajwani, Khaled M. Elbassioni, Sathish Govindarajan, and Saurabh Ray. Conflict-free coloring for rectangle ranges using $(n^{.382})$ colors. In Gibbons and Scheideler [GS07], pages 181–187.

[AFM05] Bernardo M. Ábrego and Silvia Fernández-Merchant. A lower bound for the rectilinear crossing number. *Graphs and Combinatorics*, 21(3):293–300, 2005.

[ÁFM07] Bernardo M. Ábrego and Silvia Fernández-Merchant. Geometric drawings of $k_n$ with few crossings. *J. Comb. Theory, Ser. A*, 114(2):373–379, 2007.

[ÁFMLS08] Bernardo M. Ábrego, Silvia Fernández-Merchant, Jesús Leaños, and Gelasio Salazar. A central approach to bound the number of crossings in a generalized configuration. *Electronic Notes in Discrete Mathematics*, 30:273–278, 2008.

[AG86] Noga Alon and Ervin Györi. The number of small semispaces of a finite set of points in the plane. *J. Comb. Theory, Ser. A*, 41(1):154–157, 1986.

[AGOR07] Oswin Aichholzer, Jesús García, David Orden, and Pedro Ramos. New lower bounds for the number of $(\leq k)$-edges and the rectilinear crossing number of $\mathrm{k_n}$. *Discrete & Computational Geometry*, 38(1):1–14, 2007.

[AGOR09] Oswin Aichholzer, Jesús García, David Orden, and Pedro Ramos. New results on lower bounds for the number of $(\leq k)$-facets. *European Journal of Combinatorics*, 30(7):1568–1574, 2009.

[AH89] Kenneth Appel and Wolfgang Haken. *Every planar map is four colorable*, volume 98 of *Contemporary Mathematics*. American Mathematical Society, Providence, RI, 1989. With the collaboration of J. Koch.

[AIUU96] Jin Akiyama, Y. Ishigami, Masatsugu Urabe, and Jorge Urrutia. On circles containing the maximum number of points. *Discrete Mathematics*, 151(1-3):15–18, 1996.

[Alo92] Noga Alon. Choice numbers of graphs: a probabilistic approach. *Combinatorics, Probability & Computing*, 1:107–114, 1992.

[Alo93] Noga Alon. Restricted colorings of graphs. In *Surveys in combinatorics, 1993 (Keele)*, volume 187 of *London Math. Soc. Lecture Note Ser.*, pages 1–33. Cambridge Univ. Press, Cambridge, 1993.

[AS08] Noga Alon and Shakhar Smorodinsky. Conflict-free colorings of shallow discs. *Int. J. Comput. Geometry Appl.*, 18(6):599–604, 2008.

[ASS89] Pankaj K. Agarwal, Micha Sharir, and Peter W. Shor. Sharp upper and lower bounds on the length of general Davenport-Schinzel sequences. *J. Comb. Theory, Ser. A*, 52(2):228–274, 1989.

[ASW08] Pankaj K. Agarwal, Micha Sharir, and Emo Welzl. Algorithms for center and tverberg points. *ACM Transactions on Algorithms*, 5(1), 2008.

[AvHK+03] Karen Aardal, Stan P. M. van Hoesel, Arie M. C. A. Koster, Carlo Mannino, and Antonio Sassano. Models and solution techniques for frequency assignment problems. *4OR*, 1(4):261–317, 2003.

[AvHK+07] Karen Aardal, Stan P. M. van Hoesel, Arie M. C. A. Koster, Carlo Mannino, and Antonio Sassano. Models and solution techniques for frequency assignment problems. *Annals OR*, 153(1):79–129, 2007.

[AW03] Artur Andrzejak and Emo Welzl. In between *k*-sets, *j*-facets, and *i*-faces: *(i , j)*-partitions. *Discrete & Computational Geometry*, 29(1):105–131, 2003.

[BFL90] Imre Bárány, Zoltán Füredi, and László Lovász. On the number of halving planes. *Combinatorica*, 10(2):175–183, 1990.

[BMZ09] Pavle V. M. Blagojević, Benjamin Matschke, and Günter M. Ziegler. Optimal bounds for the colored tverberg problem. *arXiv*, 0910.4987, 2009.

[BNCOS07a] Amotz Bar-Noy, Panagiotis Cheilaris, Svetlana Olonetsky, and Shakhar Smorodinsky. Online conflict-free colorings for hypergraphs. In Lars Arge, Christian Cachin, Tomasz Jurdzinski, and Andrzej Tarlecki, editors, *ICALP*, volume 4596 of *Lecture Notes in Computer Science*, pages 219–230. Springer, 2007.

[BNCOS07b] Amotz Bar-Noy, Panagiotis Cheilaris, Svetlana Olonetsky, and Shakhar Smorodinsky. Weakening the online adversary just enough to get optimal conflict-free colorings for intervals. In Gibbons and Scheideler [GS07], pages 194–195.

[BNCOS10] Amotz Bar-Noy, Panagiotis Cheilaris, Svetlana Olonetsky, and Shakhar Smorodinsky. Online conflict-free colouring for hypergraphs. *Combinatorics, Probability & Computing*, 19(4):493–516, 2010.

[BNCS06] Amotz Bar-Noy, Panagiotis Cheilaris, and Shakhar Smorodinsky. Conflict-free coloring for intervals: from offline to online. In Phillip B. Gibbons and Uzi Vishkin, editors, *SPAA*, pages 128–137. ACM, 2006.

[BNCS08] Amotz Bar-Noy, Panagiotis Cheilaris, and Shakhar Smorodinsky. Deterministic conflict-free coloring for intervals: From offline to online. *ACM Transactions on Algorithms*, 4(4), 2008.

[BS04] József Balogh and Gelasio Salazar. Improved bounds for the number of (<=k)-sets, convex quadrilaterals, and the rectilinear crossing number of $k_n$. In János Pach, editor, *Graph Drawing*, volume 3383 of *Lecture Notes in Computer Science*, pages 25–35. Springer, 2004.

[BS06] József Balogh and Gelasio Salazar. k-sets, convex quadrilaterals, and the rectilinear crossing number of $k_n$. *Discrete & Computational Geometry*, 35(4):671–690, 2006.

[BSSU89] Imre Bárány, James H. Schmerl, Stuart J. Sidney, and Jorge Urrutia. A combinatorial result about points and balls in euclidean space. *Discrete & Computational Geometry*, 4:259–262, 1989.

[Car07] Constantin Carathéodory. Über den Variabilitätsbereich der Koeffizienten von Potenzreihen, die gegebene Werte nicht annehmen. *Math. Ann.*, 64(1):95–115, 1907.

[CCHT08] Joseph Wun-Tat Chan, Francis Y. L. Chin, Xiangyu Hong, and Hing-Fung Ting. Dynamic offline conflict-free coloring for unit disks. In Evripidis Bampis and Martin Skutella, editors, *WAOA*, volume 5426 of *Lecture Notes in Computer Science*, pages 241–252. Springer, 2008.

[CEM+93] Kenneth L. Clarkson, David Eppstein, Gary L. Miller, Carl Sturtivant, and Shang-Hua Teng. Approximating center points with iterated radon points. In *Symposium on Computational Geometry*, pages 91–98, 1993.

[CEM+96] Kenneth L. Clarkson, David Eppstein, Gary L. Miller, Carl Sturtivant, and Shang-Hua Teng. Approximating center points with iterative radon points. *Int. J. Comput. Geometry Appl.*, 6(3):357–377, 1996.

[CFK+07] Ke Chen, Amos Fiat, Haim Kaplan, Meital Levy, Jiří Matoušek, Elchanan Mossel, János Pach, Micha Sharir, Shakhar Smorodinsky, Uli Wagner, and Emo Welzl. Online conflict-free coloring for intervals. *SIAM J. Comput.*, 36(5):1342–1359, 2007.

[Cha04] Timothy M. Chan. An optimal randomized algorithm for maximum tukey depth. In J. Ian Munro, editor, *SODA*, pages 430–436. SIAM, 2004.

[Che09] Panagiotis Cheilaris. *Conflict-free coloring.* PhD thesis, City University of New York, 2009.

[CKP10] Panagiotis Cheilaris, Balázs Keszegh, and Dömötör Pálvölgyi. Unique-maximum and conflict-free colorings for hypergraphs and tree graphs. *arXiv*, 1002.4210, 2010.

[CKS09] Ke Chen, Haim Kaplan, and Micha Sharir. Online conflict-free coloring for halfplanes, congruent disks, and axis-parallel rectangles. *ACM Transactions on Algorithms*, 5(2), 2009.

[CPST09] Xiaomin Chen, János Pach, Mario Szegedy, and Gábor Tardos. Delaunay graphs of point sets in the plane with respect to axis-parallel rectangles. *Random Struct. Algorithms*, 34(1):11–23, 2009.

[CS89] Kenneth L. Clarkson and Peter W. Shor. Applications of random sampling in computational geometry, ii. *Discrete & Computational Geometry*, 4(5):387–421, 1989.

[CS10] Panagiotis Cheilaris and Shakhar Smorodinsky. Choosability in geometric hypergraphs. *arXiv*, 1005.5520v2, 2010.

[CSS10] Panagiotis Cheilaris, Shakhar Smorodinsky, and Marek Sulovský. The potential to improve the choice: list conflict-free coloring for geometric hypergraphs. *arXiv*, 1005.5520v3, 2010.

[CT10] Panagiotis Cheilaris and Géza Tóth. Graph unique-maximum and conflict-free colorings. In Tiziana Calamoneri and Josep Díaz, editors, *CIAC*, volume 6078 of *Lecture Notes in Computer Science*, pages 143–154. Springer, 2010.

[dB08] Mark de Berg. Improved bounds on the union complexity of fat objects. *Discrete & Computational Geometry*, 40(1):127–140, 2008.

[dB10] Mark de Berg. Better bounds on the union complexity of locally fat objects. In Jack Snoeyink, Mark de Berg, Joseph S. B. Mitchell, Günter Rote, and Monique Teillaud, editors, *Symposium on Computational Geometry*, pages 39–47. ACM, 2010.

[dBCvKO08] Mark de Berg, Otfried Cheong, Marc van Kreveld, and Mark Overmars. *Computational geometry.* Springer-Verlag, Berlin, third edition, 2008. Algorithms and applications.

[DE94] Tamal K. Dey and Herbert Edelsbrunner. Counting triangle crossings and halving planes. *Discrete & Computational Geometry*, 12:231–289, 1994.

[Dey98] Tamal K. Dey. Improved bounds for planar k-sets and related problems. *Discrete & Computational Geometry*, 19(3):373–382, 1998.

[Dji82] Hristo Nicolov Djidjev. On the problem of partitioning planar graphs. *SIAM J. Algebraic Discrete Methods*, 3(2):229–240, 1982.

[DKKM94] Jitender S. Deogun, Ton Kloks, Dieter Kratsch, and Haiko Müller. On vertex ranking for permutations and other graphs. In Patrice Enjalbert, Ernst W. Mayr, and Klaus W. Wagner, editors, *STACS*, volume 775 of *Lecture Notes in Computer Science*, pages 747–758. Springer, 1994.

[DS65] Harold Davenport and Andrzej Schinzel. A combinatorial problem connected with differential equations. *Amer. J. Math.*, 87:684–694, 1965.

[EAS11] Esther Ezra, Boris Aronov, and Micha Sharir. Improved bound for the union of fat triangles. In Dana Randall, editor, *SODA*, pages 1778–1785. SIAM, 2011.

[Efr05] Alon Efrat. The complexity of the union of (alpha, beta)-covered objects. *SIAM J. Comput.*, 34(4):775–787, 2005.

[EHSS89] Herbert Edelsbrunner, Nany Hasan, Raimund Seidel, and Xiao J. Shen. Circles through two points that always enclose many points. *Geometriae Dedicata*, 32(1):1–12, 1989.

[ELRS03] Guy Even, Zvi Lotker, Dana Ron, and Shakhar Smorodinsky. Conflict-free colorings of simple geometric regions with applications to frequency assignment in cellular networks. *SIAM J. Comput.*, 33(1):94–136, 2003.

[ELSS73] Paul Erdős, László Lovász, A. Simmons, and E. G. Straus. Dissection graphs of planar point sets. In *A survey of combinatorial theory (Proc. Internat. Sympos., Colorado State Univ., Fort Collins, Colo., 1971)*, pages 139–149. North-Holland, Amsterdam, 1973.

[ERT80] Paul Erdős, Arthur L. Rubin, and Herbert Taylor. Choosability in graphs. In *Proceedings of the West Coast Conference on Combinatorics, Graph Theory and Computing (Humboldt State Univ., Arcata, Calif., 1979)*, Congress. Numer., XXVI, pages 125–157, Winnipeg, Man., 1980. Utilitas Math.

[ES00] Alon Efrat and Micha Sharir. On the complexity of the union of fat convex objects in the plane. *Discrete & Computational Geometry*, 23(2):171–189, 2000.

[EW85] Herbert Edelsbrunner and Emo Welzl. On the number of line separations of a finite set in the plane. *J. Comb. Theory, Ser. A*, 38(1):15–29, 1985.

[FLM$^+$05] Amos Fiat, Meital Levy, Jiří Matoušek, Elchanan Mossel, János Pach, Micha Sharir, Shakhar Smorodinsky, Uli Wagner, and Emo Welzl. Online conflict-free coloring for intervals. In *SODA*, pages 545–554. SIAM, 2005.

[Flo33] A. Flores. Über die Existenz $n$-dimensionaler Komplexe, die nicht in den $R_{2n}$ topologisch einbettbar sind. *Ergebnisse math. Kolloquium Wien*, 5:17–24, 1933.

[GP84] Jacob E. Goodman and Richard Pollack. On the number of k-subsets of a set of n points in the plane. *J. Comb. Theory, Ser. A*, 36(1):101–104, 1984.

[GS07] Phillip B. Gibbons and Christian Scheideler, editors. *SPAA 2007: Proceedings of the 19th Annual ACM Symposium on Parallel Algorithms and Architectures, San Diego, California, USA, June 9-11, 2007.* ACM, 2007.

[Hay89] Ryan Hayward. A note on the circle containment problem. *Discrete & Computational Geometry*, 4:263–264, 1989.

[Hel30] Eduard Helly. Über Systeme von abgeschlossenen Mengen mit gemeinschaftlichen Punkten. *Monatsh. Math. Phys.*, 37(1):281–302, 1930.

[HKS10] Elad Horev, Roi Krakovski, and Shakhar Smorodinsky. Conflict-free coloring made stronger. In Haim Kaplan, editor, *SWAT*, volume 6139 of *Lecture Notes in Computer Science*, pages 105–117. Springer, 2010.

[HPS03] Sariel Har-Peled and Shakhar Smorodinsky. On conflict-free coloring of points and simple regions in the plane. In *Symposium on Computational Geometry*, pages 114–123. ACM, 2003.

[HPS05] Sariel Har-Peled and Shakhar Smorodinsky. Conflict-free coloring of points and simple regions in the plane. *Discrete & Computational Geometry*, 34(1):47–70, 2005.

[HRW89] Ryan Hayward, David Rappaport, and Rephael Wenger. Some extremal results on circles containing points. *Discrete & Computational Geometry*, 4(3):253–258, 1989.

[HS84] Sergiu Hart and Micha Sharir. Nonlinearity of Davenport-Schinzel sequences and of a generalized path compression scheme. In *FOCS*, pages 313–319. IEEE, 1984.

[HS86] Sergiu Hart and Micha Sharir. Nonlinearity of Davenport-Schinzel sequences and of generalized path compression schemes. *Combinatorica*, 6(2):151–178, 1986.

[IRV88] Ananth V. Iyer, H. Donald Ratliff, and Gopalakrishnan Vijayan. Optimal node ranking of trees. *Inf. Process. Lett.*, 28(5):225–229, 1988.

[KLPS86] Klara Kedem, Ron Livne, János Pach, and Micha Sharir. On the union of jordan regions and collision-free translational motion amidst polygonal obstacles. *Discrete & Computational Geometry*, 1:59–70, 1986.

[KMS95] Meir Katchalski, William McCuaig, and Suzanne M. Seager. Ordered colourings. *Discrete Mathematics*, 142(1-3):141–154, 1995.

[KV01] Michael Krivelevich and Van H. Vu. Choosability in random hypergraphs. *J. Comb. Theory, Ser. B*, 83(2):241–257, 2001.

[Lee91] Carl W. Lee. Winding numbers and the generalized lower-bound conjecture. In *Discrete and computational geometry (New Brunswick, NJ, 1989/1990)*, volume 6 of *DIMACS Ser. Discrete Math. Theoret. Comput. Sci.*, pages 209–219. Amer. Math. Soc., Providence, RI, 1991.

[Lei80] Charles E. Leiserson. Area-efficient graph layouts (for vlsi). In *FOCS*, pages 270–281. IEEE, 1980.

[Lei84] Frank Thomson Leighton. New lower bound techniques for vlsi. *Mathematical Systems Theory*, 17(1):47–70, 1984.

[Liu86] Joseph W. H. Liu. Computational models and task scheduling for parallel sparse Cholesky factorization. *Parallel Computing*, 3(4):327–342, 1986.

[Lov72] László Lovász. On the number of halving lines. *Ann. Univ. Sci. Budapest. Eötvös Sect. Math.*, 14:107–108, 1972.

[LT80] Richard J. Lipton and Robert Endre Tarjan. Applications of a planar separator theorem. *SIAM J. Comput.*, 9(3):615–627, 1980.

[LTP09] Nissan Lev-Tov and David Peleg. Conflict-free coloring of unit disks. *Discrete Applied Mathematics*, 157(7):1521–1532, 2009.

[LVWW04] László Lovász, Katalin Vesztergombi, Uli Wagner, and Emo Welzl. Convex quadrilaterals and $k$-sets. In *Towards a theory of geometric graphs*, volume 342 of *Contemp. Math.*, pages 139–148. Amer. Math. Soc., Providence, RI, 2004.

[Mat02] Jiří Matoušek. *Lectures on Discrete Geometry*. Springer-Verlag New York, Inc., Secaucus, NJ, USA, 2002.

[McM70] Peter McMullen. The maximum numbers of faces of a convex polytope. *Mathematika*, 17:179–184, 1970.

[Met70] B. H. Metzger. Spectrum management technique. Presentation at 38th National ORSA meeting (Detroit, MI), 1970.

[MPS+94] Jiří Matoušek, János Pach, Micha Sharir, Shmuel Sifrony, and Emo Welzl. Fat triangles determine linearly many holes. *SIAM J. Comput.*, 23(1):154–169, 1994.

[MS09] Gary L. Miller and Donald Sheehy. Approximate center points with proofs. In John Hershberger and Efi Fogel, editors, *Symposium on Computational Geometry*, pages 153–158. ACM, 2009.

[MS10] Gary L. Miller and Donald Sheehy. Approximate centerpoints with proofs. *Comput. Geom.*, 43(8):647–654, 2010.

[MSSW06] Jiří Matoušek, Micha Sharir, Shakhar Smorodinsky, and Uli Wagner. $k$-sets in four dimensions. *Discrete & Computational Geometry*, 35(2):177–191, 2006.

[Niv10] Gabriel Nivasch. Improved bounds and new techniques for davenport–schinzel sequences and their generalizations. *J. ACM*, 57(3), 2010.

[NLU88] Victor Neumann-Lara and Jorge Urrutia. A combinatorial result on points and circles on the plane. *Discrete Mathematics*, 69(2):173–178, 1988.

[Pec85] G. W. Peck. On 'k-sets' in the plane. *Discrete Mathematics*, 56(1):73–74, 1985.

[PSS92] János Pach, William L. Steiger, and Endre Szemerédi. An upper bound on the number of planar k-sets. *Discrete & Computational Geometry*, 7:109–123, 1992.

[PT03] János Pach and Géza Tóth. Conflict-free colorings. Aronov, Boris (ed.) et al., Discrete and computational geometry. The Goodman-Pollack Festschrift. Berlin: Springer. Algorithms Comb. 25, 665-671 (2003)., 2003.

[PT09] János Pach and Gábor Tardos. Conflict-free colourings of graphs and hypergraphs. *Combinatorics, Probability & Computing*, 18(5):819–834, 2009.

[PT10] János Pach and Gábor Tardos. Coloring axis-parallel rectangles. *J. Comb. Theory, Ser. A*, 117(6):776–782, 2010.

[Rad21] Johann Radon. Mengen konvexer Körper, die einen gemeinsamen Punkt enthalten. *Math. Ann.*, 83(1-2):113–115, 1921.

[RSST96] Neil Robertson, Daniel P. Sanders, Paul Seymour, and Robin Thomas. A new proof of the four-colour theorem. *Electron. Res. Announc. Amer. Math. Soc.*, 2(1):17–25 (electronic), 1996.

[RV09] Pedro A. Ramos and Raquel Viaña. Depth of segments and circles through points enclosing many points: a note. *Computational Geometry*, 42(4):338–341, 2009.

[Sar92] Karanbir S. Sarkaria. Tverberg's theorem via number fields. *Israel J. Math.*, 79(2-3):317–320, 1992.

[SDG92] Arunabha Sen, Haiyong Deng, and Sumanta Guha. On a graph partition problem with application to VLSI layout. *Inf. Process. Lett.*, 43(2):87–94, 1992.

[Sha01] Micha Sharir. The Clarkson-Shor technique revisited and extended. In *Symposium on Computational Geometry*, pages 252–256, 2001.

[Sha03] Micha Sharir. The Clarkson-Shor technique revisited and extended. *Combinatorics, Probability & Computing*, 12(2), 2003.

[Smo03] Shakhar Smorodinsky. *Combinatorial Problems in Computational Geometry.* PhD thesis, School of Computer Science, Tel Aviv University, 2003.

[Smo06] Shakhar Smorodinsky. On the chromatic number of some geometric hypergraphs. In *SODA*, pages 316–323. ACM Press, 2006.

[Smo07] Shakhar Smorodinsky. On the chromatic number of geometric hypergraphs. *SIAM J. Discrete Math.*, 21(3):676–687, 2007.

[Smo08] Shakhar Smorodinsky. A note on the online first-fit algorithm for coloring $k$-inductive graphs. *Inf. Process. Lett.*, 109(1):44–45, 2008.

[SS04] Shakhar Smorodinsky and Micha Sharir. Selecting points that are heavily covered by pseudo-circles, spheres or rectangles. *Combinatorics, Probability & Computing*, 13(3):389–411, 2004.

[SST00] Micha Sharir, Shakhar Smorodinsky, and Gábor Tardos. An improved bound for k-sets in three dimensions. In *Symposium on Computational Geometry*, pages 43–49, 2000.

[SST01] Micha Sharir, Shakhar Smorodinsky, and Gábor Tardos. An improved bound for k-sets in three dimensions. *Discrete & Computational Geometry*, 26(2):195–204, 2001.

[SSW08] Shakhar Smorodinsky, Marek Sulovský, and Uli Wagner. On center regions and balls containing many points. In Xiaodong Hu and Jie Wang, editors, *COCOON*, volume 5092 of *Lecture Notes in Computer Science*, pages 363–373. Springer, 2008.

[SW10] Marek Sulovský and Uli Wagner. k-sets and continuous motion in $R^3$. In *CCCG*, pages 47–50, 2010.

[Ten91] Shang-Hua Teng. *Points, spheres, and separators: A unified geometric approach to graph partitioning.* PhD thesis, School of Computer Science, Carnegie Mellon University, 1991.

[Tho94] Carsten Thomassen. Every planar graph is 5-choosable. *J. Comb. Theory, Ser. B*, 62(1):180–181, 1994.

[Tót00] Géza Tóth. Point sets with many $k$-sets. In *Symposium on Computational Geometry*, pages 37–42, 2000.

[Tve66] Helge Tverberg. A generalization of Radon's theorem. *J. London Math. Soc.*, 41:123–128, 1966.

[Tve81] Helge Tverberg. A generalization of Radon's theorem. II. *Bull. Austral. Math. Soc.*, 24(3):321–325, 1981.

[Viz76] Vadim G. Vizing. Coloring the vertices of a graph in prescribed colors. *Diskret. Analiz*, 101(29 Metody Diskret. Anal. v Teorii Kodov i Shem):3–10, 1976.

[vK32] Egbert R. van Kampen. Komplexe in Euklidischen Räumen. *Abhandlungen Hamburg*, 9:72–78, 1932.

[Wag06] Uli Wagner. On a geometric generalization of the upper bound theorem. In *FOCS*, pages 635–645. IEEE Computer Society, 2006.

[Wag08] Uli Wagner. $k$-sets and $k$-facets. In *Surveys on discrete and computational geometry*, volume 453 of *Contemp. Math.*, pages 443–513. Amer. Math. Soc., Providence, RI, 2008.

[Wel01] Emo Welzl. Entering and leaving k-facets. *Discrete & Computational Geometry*, 25(3):351–364, 2001.

[ŽV92] Rade T. Živaljević and Sinisa T. Vrećica. The colored tverberg's problem and complexes of injective functions. *J. Comb. Theory, Ser. A*, 61(2):309–318, 1992.

# Glossary

$\bigcirc_d(n)$ minimum number of convex crossings in a set of $n$ points in $\mathbb{R}^d$. 93

$\chi_{\mathbf{cf}}(H)$ conflict-free chromatic number of the hypergraph $H$. 108

$\chi(H)$ chromatic number of the hypergraph $H$. 108

$\chi_{\mathbf{rb}}(H)$ rainbow chromatic number of the hypergraph $H$. 108

$\chi_{\mathbf{um}}(H)$ unique-maximum chromatic number of the hypergraph $H$. 108

$\mathcal{A}$ the family of all closed axis-parallel rectangles in the plane. 110

$\mathcal{B}$ the family of all closed balls in $\mathbb{R}^d$. 16

$\mathrm{b}_d(n)$ value of the ball depth problem for $n$ points in dimension $d$. 20

$C_r^{\mathcal{R}}(P)$ an $r$-centre of the set $P$ with respect to $\mathcal{R}$. 18

$ch_{\mathbf{cf}}(H)$ conflict-free choice number of the hypergraph $H$. 122

$ch(H)$ choice number of the hypergraph $H$. 122

$ch_{\mathbf{rb}}(H)$ rainbow choice number of the hypergraph $H$. 122

$ch_{\mathbf{um}}(H)$ unique-maximum choice number of the hypergraph $H$. 122

$\mathrm{cr}_k(P)$ number of crossings between the $k$-edges in the set $P$. 70

$\overline{\mathrm{cr}}_d(n)$ minimum number of simplicial crossings in a set of $n$ points in $\mathbb{R}^d$. 93

$\mathcal{D}$ the family of all closed discs in the plane. 110

$\deg_k(q_1 \dots q_{d-1})$ half the number of $k$-facets incident to the simplex $q_1 \dots q_{d-1}$. 69

$\text{depth}_P^{\mathcal{R}}(A)$ $\mathcal{R}$-depth of the set $A$ with respect to $P$. 16

$e_k(P)$ the number of $k$-facets of the point set $P$. 68

$e_{\leq k}(P)$ the number of $\leq k$-facets of the point set $P$. 68

$e_{<k}(P)$ the number of $< k$-facets of the point set $P$. 68

$G(H)$ the Delaunay graph of the hypergraph $H$. 111

$\mathcal{H}$ the family of all closed half-planes in the plane or half-spaces in $\mathbb{R}^d$. 16, 110

$\mathcal{H}^+$ the family of all closed upper half-planes in the plane or upper half-spaces in $\mathbb{R}^d$. 16

$\mathcal{H}^-$ the family of all closed lower half-planes in the plane or lower half-spaces in $\mathbb{R}^d$. 16

$\text{h}_d(n)$ value of the half-space depth problem for $n$ points in dimension $d$. 22

$H_G^{\mathbf{path}}$ path hypergraph of the graph $G$. 109

$H_n$ the discrete interval hypergraph on $n$ vertices. 110

$H(\mathcal{R})$ hypergraph induced by the regions $\mathcal{R}$. 109

$H_{\mathcal{R}}(P)$ hypergraph induced by a point set $P$ with respect to regions $\mathcal{R}$. 109

$H[V']$ the subhypergraph of $H$ induced by $V'$. 106

$\mathcal{I}$ the family of all closed intervals on $\mathbb{R}$. 109

$\text{lh}_d(n)$ value of the lower half-space depth problem for $n$ points in dimension $d$. 20

$\text{lr}_d(n)$ value of the stabbing lower centre region problem for $n$ points in dimension $d$. 22

$\text{p}_d(n)$ value of the intersecting partition problem for $n$ points in dimension $d$. 22

$\text{pcr}_k(P)$ number of pinched crossings between the $k$-facets in the set $P$. 70

$\mathrm{r}_d(n)$ value of the stabbing centre region problem for $n$ points in dimension $d$. 22

$\mathrm{s}_d(n)$ value of the pinning simplices problem for $n$ points in dimension $d$. 22

# Curriculum vitae

Marek Sulovský
born on June 24, 1983
in Brno, Czechoslovakia

1997 – 2001
**High school**
Gymnázium Brno, třída Kapitána Jaroše 14
Brno, Czech Republic

2001 – 2007
**Magistr**
Charles University in Prague
Prague, Czech Republic
Discrete Mathematics and Combinatorial Optimization

2005 – 2006
**Master of Science**
Vrije Universiteit Amsterdam
Amsterdam, The Netherlands
Computer Science

2007 – 2011
**Doctor of Sciences**
ETH Zurich
Zürich, Switzerland
Theoretical Computer Science